Conservation of Earth Structures

Butterworth-Heinemann Series in Conservation and Museology

Series Editors: *Arts and Archaeology*

Andrew Oddy
British Museum, London

Architecture

Derek Linstrum
Formerly Institute of Advanced Architectural Studies, University of York

US Executive Editor: **Norbert S. Baer**
New York University, Conservation Center of the Institute of Fine Arts

Consultants: **Sir Bernard Feilden**

David Bomford
National Gallery, London

C.V. Horie
Manchester Museum, University of Manchester

Colin Pearson
Canberra College of Advanced Education

Sarah Staniforth
National Trust, London

Published titles: Artists' Pigments c.1600–1835, 2nd Edition (Harley)
Care and Conservation of Geological Material (Howie)
Care and Conservation of Palaeontological Material (Collins)
Chemical Principles of Textile Conservation (Tímár-Balázsy, Eastop)
Conservation and Exhibitions (Stolow)
Conservation and Restoration of Ceramics (Buys, Oakley)
Conservation and Restoration of Works of Art and Antiquities (Kühn)
Conservation of Building and Decorative Stone, combined paperback edition (Ashurst, Dimes)
Conservation of Glass (Newton, Davison)
Conservation of Historic Buildings (Feilden)
Conservation of Library and Archive Materials and the Graphic Arts (Petherbridge)
Conservation of Manuscripts and Painting of South-east Asia (Agrawal)
Conservation of Marine Archaeological Objects (Pearson)
Conservation of Wall Paintings (Mora, Mora, Philippot)
Historic Floors: Their History and Conservation (Fawcett)
The Museum Environment, 2nd Edition (Thomson)
The Organic Chemistry of Museum Objects, 2nd Edition (Mills, White)
Radiography of Cultural Material (Lang, Middleton)
The Textile Conservator's Manual, 2nd Edition (Landi)

Related titles: Digital Collections (Keene)
Laser Cleaning in Conservation (Cooper)
Lighting Historic Buildings (Phillips)
Manual of Curatorship, 2nd edition (Thompson)
Manual of Heritage Management (Harrison)
Materials for Conservation (Horie)
Metal Plating and Patination (Niece, Craddock)
Museum Documentation Systems (Light)
Touring Exhibitions (Sixsmith)

Conservation of Earth Structures

John Warren RIBA, FRTPI, FSA

OXFORD AUCKLAND BOSTON JOHANNESBURG MELBOURNE NEW DELHI

For Judith

691·4 WAR

Butterworth-Heinemann
Linacre House, Jordan Hill, Oxford OX2 8DP
225 Wildwood Avenue, Woburn, MA 01801-2041
A division of Reed Educational and Professional Publishing Ltd

 A member of the Reed Elsevier plc group

First published 1999

British Library Cataloguing in Publication Data
A catalogue record for this book is available from the British Library

ISBN 0 7506 4191 6

Library of Congress Cataloguing in Publication Data
A catalogue record for this book is available from the Library of Congress

Composition by Scribe Design, Gillingham, Kent
Printed and bound in Great Britain by The Bath Press, Bath

Contents

Series Editors' Preface

The conservation of artefacts and buildings has a long history, but the positive emergence of conservation as a profession can be said to date from the foundation of the International Institute for the Conservation of Museum Objects (IIC) in 1950 (the last two words of the title being later changed to Historic and Artistic Works) and the appearance soon after in 1952 of its journal *Studies in Conservation*. The role of the conservator as distinct from those of the restorer and the scientist had been emerging during the 1930s with a focal point in the Fogg Art Museum, Harvard University, which published the precursor to *Studies in Conservation, Technical Studies in the Field of the Fine Arts* (1932–42).

UNESCO, through its Cultural Heritage Division and its publications, had always taken a positive role in conservation and the foundation, under its auspices, of the International Centre for the Study of the Preservation and the Restoration of Cultural Property (ICCROM), in Rome, was a further advance. The Centre was established in 1959 with the aims of advising internationally on conservation problems, co-ordinating conservation activators and establishing standards of training courses.

A significant confirmation of professional progress was the transformation at New York in 1966 of the two committees of the International Council of Museums (ICOM), one curatorial on the Care of Paintings (founded in 1949) and the other mainly scientific (founded in the mid-1950s), into the ICOM Committee for Conservation.

Following the Second International Congress of Architects in Venice in 1964 when the Venice Charter was promulgated, the International Council of Monuments and Sites (ICOMOS) was set up in 1965 to deal with archaeological, architectural and town planning questions, to schedule monuments and sites and to monitor relevant legislation. From the early 1960s onwards, international congresses (and the literature emerging from them) held by IIC, ICOM, ICOMOS and ICCROM not only advanced the subject in its various technical specializations but also emphasized the cohesion of conservators and their subject as an interdisciplinary profession.

The use of the term *Conservation* in the title of this series refers to the whole subject of the care and treatment of valuable artefacts, both movable and immovable, but within the discipline conservation has a meaning which is distinct from that of restoration. *Conservation* used in this specialized sense has two aspects: first, the control of the environment to minimize the decay of artefacts and materials; and, second, their treatment to arrest decay and to stabilize them where possible against further deterioration. Restoration is the continuation of the latter process, when conservation treatment is thought to be insufficient, to the extent of reinstating an object, without falsification, to a condition in which it can be exhibited.

In the field of conservation conflicts of values on aesthetic, historical, or technical grounds are often inevitable. Rival attitudes and methods inevitably arise in a subject which is still developing and at the core of these differences there is often a deficiency of technical knowledge. That is one of the principal *raisons d'être* of this series. In most of these matters ethical principles are the subject of much discussion, and generalizations cannot easily cover (say) buildings, furniture, easel paintings and waterlogged wooden objects.

A rigid, universally agreed principle is that all treatment should be adequately documented. There is also general agreement that structural and decorative falsification should be avoided. In addition there are three other principles which, unless there are overriding objections, it is generally agreed should be followed.

The first is the principle of the reversibility of processes, which states that a treatment should normally be such that the artefact can, if desired, be returned to its pre-treatment condition even after a long lapse of time. This principle is impossible to apply in some cases, for example where the survival of an artefact may depend upon an irreversible process. The second, intrinsic to the whole subject, is that as far as possible decayed parts of an artefact should be conserved and not replaced. The third is that the consequences of the ageing of the original materials (for example 'patina') should not normally be disguised or removed. This includes a secondary proviso that later accretions should not be retained under the false guise of natural patina.

The authors of the volumes in this series give their views on these matters, where relevant, with reference to the types of material within their scope. They take into account the differences in approach to artefacts of essentially artistic significance and to those in which the interest is primarily historical, archaeological or scientific.

The volumes are unified by a systematic and balanced presentation of theoretical and practical material with, where necessary, an objective comparison of different methods and approaches. A balance has also been maintained between the fine (and decorative) arts, archaeology and architecture in those cases where the respective branches of the subject have commong ground, for example in the treatment of stone and glass and in the control of the museum environment. Since the publication of the first volume it has been decided to include within the series related monographs and technical studies. To reflect this enlargement of its scope the series has been renamed the Butterworth–Heinemann Series in Conservation and Museology.

Though necessarily different in details of organization and treatment (to fit the particular requirements of the subject) each volume has the same general standard, which is that of such training courses as those of the University of London Institute of Archaeology, the Victoria and Albert Museum, the Conservation Center, New York University, the Institute of Advanced Architectural Studies, York, and ICCROM.

The authors have been chosen from among the acknowledged experts in each field, but as a result of the wide areas of knowledge and technique covered even by the specialized volumes in this series, in many instances multi-authorship has been necessary.

With the existence of IIC, ICOM, ICOMOS and ICCROM, the principles and practice of conservation have become as internationalized as the problems. The collaboration of Consultant Editors will help to ensure that the practices discussed in this series will be applicable throughout the world.

Preface

The purpose of this volume is to provide material for a fundamental understanding of the processes of repair and, where necessary, restitution of earth structures. The technical aspects of the study are treated from the viewpoint of a non-scientist who needs to understand the field.

The importance of this subject is rarely stated, but because so many peoples throughout the world have lived in structures of earths; have built, decorated and relished them; because their cultures have so extensively grown up around and among them, a great part of the environment of humanity has been shaped by earths and earth building. Its conservation is not merely the retention of a neglected facet of the vernacular architecture; it is bound into the core of living communities and their artistic creations.

Earth structures is a relatively recent field of study, so research into their repair and conservation is at an early stage and available literature is limited. The aim of this book is to remedy this deficiency, at least in part.

Acknowledgements

The author acknowledges the help of the following: Professor Dr Mohammad Maklya, Robert Demaus, Dr Zahir Othmar, J.L. Michaud, J.R. Harrison, Richard Hughes, The Getty Foundation, Charmian Shenton, Yang Ling, William Facey, Ray Deefholts, Lazar Sumarov, Paul Brown, Rebecca Warren, Joan Shih.

'Giv 'un a gude 'at an' a gude pair o' byutes an' ee'll
lyuke arter 'isself' (country adage).

Introduction

In the prime volume in this series on the conservation of buildings Sir Bernard Feilden writes:

> A complexity of ideas and of cultures may be said to encircle a historic building and be reflected in it.

This thought goes to the heart of building conservation. The object of the conservator must be the retention of the fabric and its consolidation for future use in the context of a sympathetic understanding of its past.

In their various forms, earth structures have sheltered human beings longer and to a greater extent than any other material – and they have disappeared on an equal scale. The giant walls of the cities of antiquity may be mere dimples in the earth from which they arose and the circumvallations of prehistoric fortresses no more than hilltop undulations. The houses of uncounted millions have vanished back into the ground as invisibly as their very bodies, and only painstaking examination shows that they ever existed. Earth is not a material of monuments in the same way as masonry, but a third of humanity – 1.5 billion people – live in buildings of earth, and whole cities are built of the material. The resultant forms are distinctive and its conservation takes on an importance by virtue of its scale and its ephemeral character.

The complete building is very much more than its earth walls, timber and finishes: it is a statement by and about those who created it, about the time of its making and about the subsequent vicissitudes that have laid upon it the patina of age, the scars of events; and it speaks of the social evolution which its adaptations and changes describe. In pleading for careful study of its authorship as well as the qualities of the building, Sir Bernard emphasizes that the conservator has a duty of intrinsic understanding before becoming committed to actions which affect the calibre and quality of the structure and its surroundings. The conservator must be guided by a full appreciation of the materials, the structure and the volumetric qualities, all of which are recordable and must be recorded. Beyond this he or she must be sensitive to the quality of surfaces and the often intangible evidence of its use and habitation. Conservators must be aware of the destruction of such evidence which repair may necessarily entail and beyond this again must perceive the status of the building, its origins, the reasons for its alterations and the function it will perform in the future. Sensitivity to such problems is not new but attitudes change. Circumstances have heightened perception: they have added to and sharpened the analytical tools at the disposal of the conservator.

Masonry and earth structures differ fundamentally. In particular, they diverge in the permanence of surface: stone and brick, which mellow, and earth, which erodes in decay. The demands upon the conservator vary accordingly. The art of conservation demands wide expertise – in the nature and materials of the building or artefact, knowledge of its place in history, an assessment of its form and function and, in the case of buildings, skill in team leadership. Throughout the process the mind

Figure I.1 The Great Wall of China – the western end at Giau Guan. Famed as man's largest historic construction, the Wall is an earth structure throughout its length, although stone-faced at its eastern end.

must be disciplined and the perception as keenly directed to the future as to the past.

Earth structures by their nature tend to be massive. Nevertheless, they succumb to the inevitable forces of destruction and of these, human beings are one of the most potent. Part of the art, therefore, is in the disposition of the building and its use by humanity. In this social dimension the conservator may be relatively powerless – at best an adviser and

facilitator, manipulating law, finding funds and securing effective utilization.

Conservation serves several purposes. In the social context the fabric of the past is meaningful in the same fundamental way as the landscape. People deprived of the environmental asset of their past lack tangible evidence of their own history. The sense of loss can result in instability and a yearning for security. The distinctive shapes of the earth

Figure I.2 The medieval earth walls of Raqqa in Syria, illustrating how whole cities vanish back into the soil from which they arose.

villages of sub-Saharan Africa provide its inhabitants with evidence of their origins in just the same way as do the mellow brick walls of some Hanseatic town whose burghers sense in their environment the deep roots of their trading past. No conservator need be ashamed of admitting that one fundamental purpose of the work is the retention of a fabric which meets the deep psychological needs of those who inherit it and pass it on.

An allied purpose of conservation is the retention of a culture. While the social purpose may be to provide continuity within the historical past, the cultural purpose is to retain the creative achievement represented by the buildings, the town or the landscape. The intellectual endeavour which will have gone into the creation of architecture can only be fully represented by that architecture itself. The level of cultural achievement may not be high but it may nevertheless be important. In the oasis villages of northern Arabia the

patterns and repetitive devices applied by tribespeople to their woodwork and to the mud which coats their buildings has a quality which makes them unique in time and in place and in that sense they are as important as the terracottas displayed on Italian Renaissance buildings.

Culture is commemorated in the artefact and the qualities that created civilizations are measured in the architecture they left behind as much as in their literature or their music. Societies may disappear but the cultural artefact remains to provide intellectual access to those who created it. There may be no direct continuity between those who retain the structure and those who made it, and at times there may be only a derivative link, but its retention is perhaps all the more important for that reason alone.

These arguments of purpose are historical and to them must be added one other – science. Any historic fabric is evidence of the

Figure I.3 A palace at Birket Muz, Oman, rendered uninhabitable by air attack, demonstrating how man is one of the most potent forces of destruction.

circumstances and techniques of its age of construction. The circumstances of its building, determined by available materials and the use of techniques inherited or invented, are the prime evidence of the movement of civilizations and one purpose of conservation must be to retain this evidence. But the artefact also encapsulates scientific information. Fragments of organic material permit dating by the measurement of atomic decay – the carbon-14 technique. Other analytical techniques include thermoluminescence, obsidian analysis and the emergent technique of optical dating by quartz, which use the effects of long burial to provide alternative methods of dating by sampling. Other techniques are known but little used, such as the qualitative measurement of saline impregnation; others have yet to be perfected or even to be discovered. The analysis of soil particles and air-borne deposits such as pollen can provide evidence of conditions and dates of construction. These and

other forms of scientific analysis provide information on sources of materials and distribution of the origins of materials which in the long future of conservation will add to our knowledge of the past. The loss or destruction of such evidence is to be avoided, particularly in view of analytical techniques yet to be invented which may extract further knowledge from even apparently unimportant artefacts.

Conservation can serve one further purpose, sometimes treated as a justification – the economic utilization of the structures. As has been discovered by many builders in the past, it can be less expensive to reuse a building than to build anew.

The adaptation of historic structures to new or enhanced usefulness is often an economic ground for conservation. Earths and brickwork are massive and involve substantial labour. Patination and the maturity of age are qualities which they gather and which cannot readily be simulated nor honestly created

Figure I.4 Detailed patterning applied to timbers carrying mud roofs in the Nejd, Kingdom of Saudi Arabia. Similar patterns are used on doors and earth walls in the vernacular building of this region.

afresh. Where society has a use for buildings of this calibre there is a strong economic ground for their conservation and skilled repair.

To describe a purpose of conservation as economic, however, raises the concern that where economic advantage cannot be demonstrated a converse argument can be used for demolition and it may be more wisely thought that the economic argument is secondary to

the fundamental purposes of historic continuity and cultural retention.

Strong in compression and weak in tension, massive by nature and generally heavy, earths and brick are, with brick and stone, the characteristic material of walling. Being inert in a practical, if not a chemical, sense, they are durable and take on special hues and characteristics with age. The methods of assembly and forms of construction have

Figure I.5 Monumental building – the minaret of the Great Mosque at Turfan, in Chinese Turkestan. Complex pattern-making on the massive stabilized brick/earth mortar minaret of the mosque.

produced a wide variety of historic characteristics intrinsic to style, age and place. Sympathetic repair requires the use of materials and techniques which are appropriate to the context and produce a result that is acceptable in terms of the historic qualities of the place or building.

In conservation the policies applicable relate to the degree of urgency, the nature of the defects, the type of threat and the intended use of the building, but in deciding on a course of action a number of overriding principles apply. Foremost among these is the

principle of **minimal intervention** coupled with **reversibility**. The ideal is that the historic fabric is interfered with as little as possible and that any intervention should be capable of being withdrawn, leaving the fabric as it was.

Conservation must always be undertaken with **integrity** – the determination that materials appropriate to the purpose are used in a fitting manner. Coupled with this must go **sympathy** to ensure that new work is consonant with the old. The character and tone of the building are determined by the

Figure I.6 Modelled clay, stabilized with lime or gypsum, used to provide decorative modelling for interior and sometimes external work, as in this case in the 10th–13th-century work in a mosque in Zuvara, Iran. Modelled decoration has been a tradition in the Middle East since pre-Muslim times, executed in stabilized earths, plaster and fired wares.

original work and any intervention must accord with the overriding qualities of the original. **Honesty** dictates that all work is what it purports to be and **datability** allows that, on analysis, the age and nature of any intervention can be proved. A further principle might be defined as **location**, to underscore the relationship of a structure to its site. It is argued that a building is irrevocably a part of its location, being created for and welded by use to its site. While there are some exceptions, in the case of buildings that can be moved, this concept of location does encapsulate an important aspect of the integral relationship of a building to its surroundings.

Fundamentally these principles aim at honesty in expressing the time and nature of the work of conservation, sympathy in ensuring that it does not obtrude, integrity in its being of the appropriate calibre and reversibil-

ity so that future conservators need not necessarily be bound by current work. To these some conservators would emphasize a minimalist approach, arguing understandably for doing only what is most essential for structural reasons, leaving undone even work necessary to control decay – a philosophy which can be described as decelerated erosion. Sometimes such principles are clearly in conflict with other objectives and for this reason conservators must understand the priorities which govern their work and be prepared to use their principles as the overriding criteria.

There can be instances where no action whatever is justified. A surviving structure below ground may be analysed by remote sensing techniques, such as magnetometry, and be left undisturbed; but most structures will require some form of repair or restoration if they are to survive.

An example might be a mud brick structure of antiquity exposed in archaeological excavations. The minimum intervention would be careful exposure and recording followed by careful reburial. Such action, however, would be unattractive to owners or authorities wishing to exploit the site for tourism. They might seek the stabilization of the remains for purposes of permanent access and this, in turn, might entail the **irreversible** introduction of chemical compounds and the **reversible** building of shelter structures for protection.

Stabilization might be more satisfactory visually than the shelter structure and, therefore, more acceptable in terms of sympathy while being less acceptable in terms of reversibility.

An acceptable proposal might be the construction on top of genuine material of a replica or reconstruction of the original building. However unacceptable this action might appear, it could be preferable to a policy of apparent non-intervention – of leaving the original structure exposed at the mercy of the elements and visitors, where such a policy could quickly result in the total destruction of the original material.

Even reburial might, however, carry with it severe disadvantages where, for instance, the pattern of drainage or vegetation has been changed by new circumstances altering the water table or allowing root penetration. Sometimes destruction is unavoidable. One archaeological layer might have to be removed in favour of the retention of another. Where major constructional works will eradicate the site, the option may be abandonment or removal of the material to an alternative location. The nature of the intervention to stabilize or secure the original construction will be determined by the circumstances and overriding imperatives but the nature of the material also plays a part.

A structure built in mud brick can be removed only with great difficulty, suffering in the process inherent damage to its nature, even though in the final event it may be restored visually. Brickwork, however, may be removed *en bloc* or piecemeal and reassembled skilfully to recreate the original arrangement, retaining the pattern of wear and weathering very precisely. In the process the mortar may be partially or largely renewed but the evidence of this renewal is no more than would have been the case had a building remained *in situ* but required repointing.

In postulating this simple series of possible events, a wide range of options become apparent and important definitions arise. They may be summarized as follows:

Deterioration, controlled or uncontrolled There is no philosophical compulsion to preserve artefacts of the past. It is accepted that a mountain should erode naturally without interference and by similar argument a ruined structure may reach a state where its contribution to the landscape is one of decay. Occasionally people are moved to counteract erosion upon a mountain (particularly if humans are the cause of it), and likewise may feel inclined to slow down or halt the erosion of a monument. Alternatively the structure may be left to decay naturally out of a policy of non-interference or simply as a result of economic necessity.

Preservation A ruined structure may be 'frozen in time' by judicious protection from further erosion. It may be stabilized in a particular state, perhaps paying special attention to the degree of ruination. It might, indeed, be encapsulated in another overriding structure – a process becoming more common as urban archaeology reveals more structures relating to urban history. By the removal of the weathering, decay can be halted without the introduction of special stabilizing techniques, but even when the structure is not encapsulated the careful use of techniques which protect the core of the structure can slow down the rate of erosion so that it proceeds very slowly.

In earth structures patently this is less easy than in masonry but more sophisticated techniques of control based on chemical and physical methods of protection will increasingly be able to protect and restrain the rate of loss.

Maintained utility This is the most common and desirable circumstance for a built structure and it presupposes that there can be a continuing function compatible with its original use. In such a case the building continues to serve a useful purpose, being maintained with special care so that the structure is not subjected to unnecessary stress, erosion or decay while generating the income which will ensure that its costs of maintenance are met. Continuous attention must be given to ensure the best possible performance and this will generally mean the maintenance and occasional replacement of building components, some of which last longer than others.

An earth structure will suffer the decay of its outer skin and consequent renewal on a cycle of high frequency. In some instances an outer protective coat, perhaps of lime wash, will be renewed even more frequently and although the use of the structure may change, its condition will be stabilized.

Maintenance involves repair. Repair is simply the making good, on a like-for-like basis, of worn material with a replacement or alternatively of the insertion of compatible material to make good elements which are lost or in the process of decay.

Other terms commonly used in the process of conservation are:

Renovation or renewal Major repair of earth requiring the removal of substantial parts of the original fabric and their replacement on an extensive basis is best described as renovation. It is likely to involve the removal and restitution of several elements simultaneously. Problems are often multiple. Rot may have caused the decay of timbers which in turn have affected the stability of mud brick. The renovation will entail the removal of that part of the structure which is no longer viable and its renewal, perhaps reusing the original materials.

The loss of part of the original material is followed by rebuilding which may include the introduction of some new material. Neither decayed timber nor removed plaster will be suitable for reuse in renovation, although the earths may be repairable. In some circumstances a decorated surface fresco painting or mosaic may be removed and replaced on a renewed substrate.

Restoration It may be necessary to restore the structure to its original condition and for this purpose later interventions may be altered or removed. These actions are sustainable only if they are part of a properly thought-through policy of conservation. Overriding policy considerations may demand structural actions involving serious physical alteration in the interests of restoring a structure to its original form, but such measures are only tolerable in a context which justifies them on comprehensive assessment.

Refurbishment Extensive renewal or modification of secondary elements of a building may be required to adapt the structure to a new purpose. The introduction of modern services is perhaps the most common instance and is normally accompanied by redecoration and often by remodelling. The enormous world stock of earth houses is, in broad terms, a stock of structures lacking modern amenities. Their introduction involves extensive refurbishment. New pipes and ducts, cables and fittings are accommodated by cutting into the historic structure which is then repaired or resurfaced. As the environmental consequences of these changes make an impact on the structure, serious deterioration can take place unless such consequences have been foreseen. Water leakage into earth structures can cause dramatic slump and structural failure and the refurbishment itself will induce new patterns of use, new loads, cooling or heating systems causing change in humidity and temperature and physical alterations which may stress an ancient fabric in ways to be foreseen and guarded against only by skill and experience.

Reconstruction The remaking of a structure by reassembly (anastylosis) is a dramatic level of intervention justifiable only by extreme decay or major calamity. A historic earth building shattered by earthquake, brought to collapse by flood, or demolished by bomb may justifiably be reconstructed. While its doors, windows and timber floors may be repaired and reinstated, the material of its earth walls can only be reused by remixing the material to form new blocks, bricks or mass walling.

Reconstruction may, nevertheless, be justified in historic terms. In essence the rebuilding of a structure in its original form using materials which survive or are identical with those originally used may be a proper part of conservation, but rarely does such an operation take place without the introduction of more modern materials or new technology. Reconstruction may be justified by external

Figure I.7 Conservation: a continuous programme of repair which renews decayed surfaces and maintains structure (courtesy Dr Paul Brown).

Figure I.8 Restoration: the city wall of Sana'a, Yemen, rebuilt using traditional methods. Architect: Myriam Olivier (courtesy Dr Paul Brown).

and internal contexts. Where the damaged building is a vital part of an urban scene or streetscape, or a focal element in a landscape, reconstruction may be preferable to demolition to preserve the established historic context. Where the contents of an interior have been saved and can be replaced, the reconstruction may be justifiable as being the best method of their display in the most meaningful context.

Re-erection or relocation The building may be removed from its original site and replaced elsewhere. The process may involve reconstruction and anastylosis or in some instances a building may be moved as a complete entity. Framed buildings and burned brick structures may be so moved; earth structures only rarely.

The essence of this process is to extract a whole or part structure from one context where it is archaeologically verifiable in order to implant it into another which may be synthetic – as in a totally new museum environment – or contextually different – as where a building is removed from one part of a town to another.

The relocation of earth structures is only possible in the most exceptional cases although earth components, such as panels of wattle and daub, are transportable and reusable, if sufficient care and skill is deployed.

A further clarification of terms is relevant generally and specifically to this work. *Bricks* and *brickwork* will be used specifically for fired brick. *Masonry* will be used for structure which is the work of masons, i.e. stonework, even where it does incorporate a small proportion of brick. *Earth structures* will include dried earths, i.e. mud bricks, and the term *adobe* will be restricted to the geographical areas to which it is pertinent, i.e. the American continents, where it means both blocks of dried earth and the raw material from which they are derived.

All historic buildings need care and skill in their maintenance and differing types of structure demand different skills and appropriate routines of care and maintenance. Both brickwork and earth structures are of such age and continuity that long traditions of maintenance routines have evolved relating to climate, locality and nature of the materials and these routines have become an established part of the lifestyle and a feature of the configuration in the buildings themselves. The projecting poles and stone corbels of walls and domes that need frequent rerendering with mud are instantly recognizable as design features in the vernacular architecture of the regions where they occur. Many more subtle features model or affect the design of buildings right through the social scale. The drip moulds which protect openings in the brickwork in wet climates have become distinctive as design features, illustrating that detailing can be as important a distinction as the material itself.

These terms are common to the whole art of building conservation and relate to work in stone masonry particularly. The debates of the first organized intellectual groups in the 19th century focused primarily on the great stone monuments of antiquity. Their voices were raised against destruction by neglect and by over-restoration.

The consequences at the time were that the dedicated endeavours of the restorers came into question and the alternatives of the care and retention of all ancient and weathered surfaces was championed as the antithesis of restoration. John Ruskin fulminated against the restorer as the great destroyer and rose in anger at the extensive repairs, renewals and restorations carried out on the church of San Marco in Venice during the latter part of his lifetime. Meanwhile the Society for the Protection of Ancient Buildings was formed in London. William Morris, its prime advocate, set out principles of care and conservation totally opposed to the unnecessary renewals of surfaces, the improvements in stylistic detail and the smartening-up of historic buildings.

These voices were at the forefront of intellectual enquiry in the conservation field and the processes they went through have been reflected in many subsequent pronouncements. The deterioration of old buildings and the ambitions of owners to bring them into a state of newness have persisted since these arguments were fully evaluated. At the one extreme stand those who create 'old' structures where none were before, whose brickwork built perhaps on ancient foundations is entirely new and may simulate the old; at the other extreme are the protagonists of the retention of any part of the existing fabric and the retention of those voids and evidences of missing elements which betray the history and events of the passage of time. The motivations and arguments are not simple but the coherence of views grows increasingly important in the face of internationally recognized standards and the growing volume of legislation.

Building conservation, like medicine, demands preventive care which is overlooked at great cost. The maintenance of sophisticated machinery on which human life depends has accustomed recent generations to the concept of planned inspections and planned maintenance and also to the idea that a professional inspectorate is the most efficient answer to the need for planned continuing care. The earliest signs of problems arising are often slight and it is necessary to differentiate between important symptoms and those which can be disregarded.

Such symptoms may be changes in the levels of ground water or loss of wind protection which might appear to have little significance but can lead to rapid failure in earth buildings. A bulge in an earth wall may appear to be nothing more than a natural contour but it may presage a significant movement following a change in loading pattern or a loss of internal integrity. The nature of cracking in brickwork will be a consequence of the distribution of loads and stress. The movement or strain will have direct implications for trained observers alerting them to the nature and point of the application of pressure, the degree of movement and the possible cause.

Plant growth and staining may indicate changes due to water content. Differential behaviour of paints may indicate problems in the substrate or surface coloration may betray problems of salinity.

Other forms of maintenance depend on assessing the nature and rate of weathering or erosion and the consequences of the failure of other building components. The effects of previous interventions are often ignored, since it is assumed that they have been effective. In practice incompatibilities in materials or parts of structure may not become apparent for some considerable time. In many instances new materials and new techniques of application demand the most perceptive use and care in monitoring. The knowledge gained from subsequent analysis of performance is the most important part of the operation. This knowledge may have to be disseminated in addition to being retained in the context of the building itself.

An essential part of the continuous care needed by historic buildings is a programme of management. Regular inspection at intervals determined by the nature of the structure must be recorded and reported. Appropriate work, perhaps scheduled well in advance, and a co-ordinated work programme carried out, will economize in labour as well as providing early responses where necessary and effective economic repair.

Historic building is, by nature and definition, long-lived, outlasting the generations who use and care for it, and therefore records of work done are of more than a passing interest to future management and historians; they become an important tool and store of knowledge in the operation of maintenance programmes. They should contain information as to the sourcing of materials, the nature and extent of work, the types of work carried out and its sequencing.

Such methodical operations will, of course, cover the whole of the building fabric, not simply the basic structures and these, being generally the less vulnerable of the components, will probably feature less in the normal maintenance programme. Nevertheless, it is helpful to future workers and conservators to know exactly what work has been carried out. Armed with this information future conservators will be able to assess the effectiveness of the conservation techniques previously used and model their work accordingly.

Sometimes action distant from a historic place can be vital to its future. Diversion of a water course or arterial road may save an earth building from a flash flood or a thin-walled structure from the vibrations of heavy traffic that will destroy it. Defensive conservation of this sort can be achieved only by foresight, knowledge and skill, coupled, perhaps, with determination and political acumen. Failing the right action at the right time, the ensuing and avoidable damage will require repairs which interfere with the historic fabric. Avoidance of loss is a measure of conservation achievement. A mud wall, standing unchanged, may not be apparent as an act of conservation, except to those who installed the drains that kept it dry, preventing collapse. It is generally true that the less visible the work of conservation, the greater the success of the conservator, providing always that the building remains sound, usable and as long-lived as can be foreseen.

Defining the point at which repair work should be undertaken is a matter of skill and judgement. The life of a historic building should be extended by simple measures such as minimizing wear, providing protection and avoiding unnecessary intervention. Redecorations should be extended to the maximum permissible cycle consistent with longevity and the judicious replacement of components will be governed by a minimalist approach. Better an original wall with a decayed face than an expensive new replica – and even better an earth wall under a newly thatched capping than the loss of its topmost course.

There remains the danger of pressures from the powers of commerce, business and politics to reflect self-pride and ambition in the restoration of monuments, resisted by an increasingly coherent conservation lobby. Between these two the judgement of the building conservator is poised. In advice and decision-making he or she becomes the arbiter and executor of a process which increasingly affects the status and well-being of a community. A decision not to repair may be as significant as a resolve to continue lime-washing an earth-walled structure or to authorize the rerendering in mud and straw of an earth wall

Figure I.9 Domed structures formed in earth block showing, in decay the structural delicacy of the semi-ellipsoid shells that characterize the vernacular building of central Syria east of Homs.

which depends on this technique for its protection. While many of the problems faced by the conservator are specific to regions or places, the ethics and aesthetics of building conservation have in common many aspects and the debates have in this century been international. Standards, charters and guidelines proliferate, calling for care, retention, responsibility and carefully applied skills.

In earth structures all this intellectual endeavour is relatively new. Consequently, the adaptation of conservation principles to mud brick, adobe and other forms of earth building is almost a fresh endeavour; much of the technical work is at the experimental stage and the philosophical approaches diverge from culture to culture.

It is abundantly clear that in the traditional domestic forms of building, so widely represented by structures of earth, is deeply embodied the culture of many peoples across the globe. If these cultures are to be retained with integrity their buildings must survive harmoniously and in large number.

1

An overview: artwork to earthwork

There is no more basic form of habitation than the hole in the ground, and this the human race shares with many other animals as the fundamental form of dwelling. Whole civilizations have flourished as troglodytes and millions of people still live comfortably in this way. The early habitation, tomb or temple may be a cave in reasonably stable ground and some of these holes have reached the state of art which deserves the term architecture and which become objects of conservation. Many of them fall into the category of building by consequence of extension outward. Often it is the products of the excavation which are converted into building materials and project the space forward. Other building materials are used as a lining and architectural skills transform the interior with colour and the exterior with bold massing and stylistic detail. Some reach high levels of sophistication and it is on the plateaux of northern China that the greatest numbers of cave dwellings now exist.

On these great uplands where rivers carve down through compacted loess (wind-blown) soils, a Buddhist culture flourished before its destruction with the coming of Islam. The adherents of that earlier, more peaceable philosophy invested labours of love and piety in their temples and housing, with statuary and panels in highly finished, painted earths on earth brick, which formed a lining to walls behind rendered earth brick facades. Pious and malicious damage, neglect and the depredations of artistic treasure hunters have still not totally defaced or destroyed these fragments of a lost civilization. Paintings still cling to the walls in their thousands, preserved in the relatively dry atmosphere of the region despite extremes of annual and diurnal temperature. These long-abandoned temples in the cliffs of the trade routes still hold giant representations of Buddhas and Bodhisattvas 30 metres in height and formed in clays made coherent in multi-layer application on wooden armatures or cores of unexcavated earth. Their highly finished surfaces ultimately were smoothed with gypsum plasters before being painted, and around them stand background panels formed and smoothed in the same way and also richly painted. By their survival, their rarity, their artistry and their historical significance such works stand high among the monuments demanding care, and basic though these structures are, they are almost supreme among the problems of conservation in earths. An important principle lies at the heart of these constructions. They carry no evidence of shrinkage, either in the panels or in the statuary, yet they contain substantial proportions of clays and these clays, like all others, shrink upon drying. In some instances the clays have been applied in bulk, the shrinkage has taken place and by a process of further application and burnishing the initial cracks have been filled to cohere into a solid panel. This remarkable effect is the product of a multi-layer application which can succeed in building up solid masses and plain surface panels showing no evidence of shrinkage cracks, although their size and surface would suffer substantial cracking in normal application.

There is no published evidence to indicate the proportion of such sculptures built up in this way, but their existence can be attributed to the laborious techniques applicable to works of art coupled with the availability of dedicated monastic labour. This special circumstance has allowed the construction of elaborate and sophisticated works of art.

1

Figure 1.1 Dunhuang, Chinese central Asia. Statuary formed on earth cores finished in painted plaster and preserved in caves in a region of low rainfall.

1.1 Early earth construction

Across the plains of the Fertile Crescent – eastern Turkey, Syria, northern Iraq and through parts of Iran to southern Russia – the landscape is studded with tells, tokens of long-established building operations. These tells, or flat-topped mounds rising from the plains, were created by the inhabitants of settled townships who continually carried in, for building purposes, earth, mortars, bricks, stone and wood; and when these decayed they were trampled underfoot to be built upon again. They were never returned to their place of origin and so the mounds are testimony to long human habitation in the region. Early

Figure 1.2 Shanxi. China. Entrance to a house cut into a solid deposit of wind-blown soil (loess).

building other than of stone or perishable materials such as skin seems to have been of clay or of mud daub on a wicker frame and the earliest bricks have tended to be of lumpy shape, settling into the plano-convex form commonly used long before the making of rigid, squared, sun-dried brick. The reason was entirely practical. A man can carry a lump of earth which is a good deal heavier than a self-supporting flat section of green or plastic clay. If it is shaped with a hump in the middle it can be carried with hands placed under it at about the quarter point without undue brick deformation, even though it remains sufficiently plastic to place in a wall. This was the primal method of construction requiring no mortar and, therefore, no tools other than a spade. The lump was placed and compacted as part of the simple continuous process requiring only one operation. It was a step of significance which led early builders to dry the lump before use. This led to the more complex process of mixing of straw and mud, its placing in moulds, drying and then the construction of walls using soft mud to provide a bedding or mortar.

Such a step was probably taken to achieve the building of arches and barrel vaults, a technique already known in stone. A plastic material such as a soft lump makes a bad vaulting material, particularly if the vault is constructed on the inclined arch principle, in which the arch leans back against an end wall and is built up progressively (pitched vaulting), the support of each succeeding voussoir being dependent upon its adhesion to the slightly inclined preceding arch against which it is built. Thus the use of sun-dried brick and the plano-convex brick or lump were parallel techniques available to the early builder and were well-established by 3000 BC, when the baking of clay had already provided humans for many centuries with pottery and votive objects.

Figure 1.3 Earth construction typical of Europe, the Balkans, through mountainous Turkey, Mesopotamia and the Caucasus to the Himalayas. Here earth block is used in the understorey with a light timber framing above filled with similar block. An outer skin of rendering – stripped from this example – may be used externally.

In parallel has run the simplest form of earth construction, the ditch and bank. These remains of human building betray fields, boundaries, defences, bases of buildings and lines of roads. Many are of high antiquity and it is a general rule that the further back they lie in time, the more significant they may be, for many of them antedate written history and very often constitute the prime visible and extant remains of early peoples. Many have become earth features in the landscape and, although deformed by subsequent agriculture, the larger of them remain dominant masses on their sites even today. Universally they have been eroded, their ditches filled and the embankments have fallen away to a natural angle of rest so that only by the most careful archaeology can the original profiles be re-established. In some cases such investigation has been done on a large scale and to retain the evidence the site has to be enclosed and protected. The evidence of the Ban Po culture in northern China now stands scraped, revealed and instructive under a major roof whose protected environment produces problems of evaporation and salt deposition. In most cases nothing more can be justified by way of archaeology than experimental or very localized excavation and the conservator's problem is less that of providing covering than of protecting the already weathered remains of the monument from further decay. Where the site is grazed and grass-covered, as so often in northern Europe, the usual view is 'leave well alone'. There are two caveats, however: damp conditions with either tree growth or with excessive wear by cattle or vehicles. Changes in the drainage regime and increased moisture content of the soil can provide a much softer ground base and, therefore, increase the erosion. Unsuitable types of grazing technique can wear down the surface very rapidly by the tread of heavy animals. Even excessive grazing by small, light animals can so erode the cover as to increase the rate of surface decay. Given a satisfactory regime, however, the best protection to large-scale earth monuments is a well-maintained turf from which rodents and other burrowing animals are deterred. Substantial trees do not fall into the category of acceptable vegetation. Deep roots are themselves a disturbance but the overturning of mature trees in storms represents the largest natural danger to such monuments, filling the ditches, breaking the banks and providing cover and space for burrowing animals. The landmark created by 19th century beechwood planting of Chanctonbury Ring in Sussex – a major prehistoric earthwork – would be unacceptable by modern standards of conservation due to the damage caused by tree growth and decay. Its replanting after the storm of 1987, however, was unavoidable because it had become so prominent a visual feature.

1.2 Forms of earth construction

Excavation and deposit is the essential process of earth construction. It commonly occurs in several ways. In the first the excavation out of which the material for earth brick is taken itself forms part of the curtilage or even the reduced floor level of the building. By extension of this method, the inner and outer ground levels are reduced, leaving the native earth as the lower part of the walls. The excavated void forms part of the construction. In the deep loess beds of northern China this was a particularly common technique in the massive constructions which characterized the major early settlements, cities and sanctuaries. In more modest English houses the excavation has been left as a void or used as a cellar under the building.

In other circumstances the excavation is the entire product, as in the hollowed-out volcanic deposits of Goreme in Turkey. Elsewhere again, the excavation is entirely separate. In India the pit from which the earth has been taken to make the village dwellings becomes the tank holding water essential to survival in the dry season.

Between these extremes of highly wrought sculptural works and the massive earth structures that have formed castles, surrounded cities, delineated countries, dammed rivers and divided seas, there has been created an immense range of walled constructions serving as homes, work places, governmental and religious buildings and for every agricultural purpose. Timber was widely used for roofing. For controlling the rise in ground water and securing foundations stone has been predominant, and to surface internal and external vertical surfaces mud plasters, lime cement, gypsum renders and lime washes have served extensively.

Earth structures, like most buildings, are multi-material constructions and the task of the conservator, therefore, is concerned with the interaction of these components. Some considerable proportion, however, of surviving earth structures consists virtually entirely of the one material, and this range extends beyond the massive constructions which in contemporary terms might be thought of as civil engineering, to humble buildings in great number, particularly in the tropical zones. The 'beehive houses' of Syria, the villages of upper Egypt, the kasbahs of Morocco, the townships on the Iranian plateau, the circular hutments of the Volta and the adobe dwellings of Mexico and New Mexico are merely some of the prime examples of a wonderfully extensive world heritage of earth construction, much of it

Figure 1.4 Chinese Central Asia: Jia Ohe (pronounced *JewHay*). Wall of the early first millennium AD in which the lowest section is native earth. On this stands the visibly less compact rammed material excavated from surrounding areas.

humble in origin, and humble in function, regarded as outmoded and unimportant by owners and users and at risk for reasons not related to its utility. It is a heritage yet to be adequately valued and by its fragility its survival depends upon maintenance which in the face of progress it is often denied.

Much the greatest part of building in earth has been carried out by four basic methods:

- the placing of prepared dried components (mud brick construction)
- the placing of earth in semi-plastic form (earth construction)
- the use of semi-liquid earths
- the compaction of earth between restraining surfaces, i.e. permanent or temporary formwork (*pisé* or rammed earth construction)

Figure 1.5 De Gau, China. Earth brick backing within a loess excavation on which gypsum-stabilized earth and ultimately gypsum plaster provides a base for mural painting.

Wet earths are used as mortars, daubs and renderings (plasters). The first type of construction always requires a mortar, which is usually a fluid mud likely to dry somewhat weaker than the brick itself. In the second form of construction the bricks, blocks or lumps may be laid without any further mud mortar, as their plasticity allows them to be shaped to their position with some compaction. In the third and fourth form of construction intermittent or continuous pressure – pounding or rolling – is an essential part of the process.

Various local descriptions and forms of nomenclature relate to specific areas and variations of technique. Of these terms one is overridingly important – the word *adobe*, which is so thoroughly established that its

Figure 1.6   Iran: medieval castle at Faraj. The collapse of a double-chambered tower demonstrates the slender cross-sections achievable with the use of squinches and secondary vaulting in the haunches of vaults. These techniques are apparent from the early Middle Ages and precede the structural magic of Persian brickwork.

continuance in compound use is unavoidable. In the southern USA, Mexico and South America, earth construction and the raw material itself are universally described as adobe, whether or not the earth components are fully dried before being applied to the building. It would be pedantic to argue the case for a change of terminology which is so deep-rooted in the local vocabulary.

In tropical and subtropical America, manufacture has reached the stage of large-scale commercial exploitation. The term *adobe* has become collective and a single brick in its dry handlable form is known as an adobe, anglicized in the plural as adobes! The root deriving from a Coptic (or earlier) word (*toob*) meaning a lump was taken into Arabic as *al dub*, transmitted to the Muslim societies of the western Mediterranean and thence by Spanish colonists to America. The word *adobe* has become so universal in the Americas – as both

noun and adjective – that its use as a verb seems imminent.

While modern commercial experience and scientific technology have now established with some precision the most advantageous soil mixes, there is no differentiation in historical construction between mixes suitable for particular types of work on the basis of structural method except, perhaps, in minor local contexts. In other words, local technique tended to be determined by tradition and recognized structural methods and the material employed would be that found most suitable in the locality. Thus, with present knowledge we might well say that a particular mixture employed as lump material would have been more suited to construction in *pisé* work, but this type of knowledge was rarely available to the vernacular builders of the past and the conservator not only needs to recognize the ancient methods of construction used but to be able to make a conscious decision as to the merits of repeating that structural form as opposed to advantageous alternative methods.

To this generalization there is one important exception. By the late 18th century the transmission of technical information by written and drawn publication had become effective even at the humble level of earth construction. In consequence, military and civil building in developed countries and their colonies displays ideas and improvements in technique transmitted by book to areas they would not have reached by example.

It was at this point that science came to bear on earth construction.

1.3 Strength and stability

In terms of composition of materials, there are generally two parallel factors affecting the strength and stability of the final earth structure. The first is the presence of a setting agent, perhaps calcite or calcitic material, often as a chalk or a lime, or the presence of gypsum or the presence of a modern cement or even the presence of a bituminous or organic binder. These materials, by crystallization or by fluidity and adhesion, can significantly affect performance in dry and wet modes by restraining the earth particles. The second and alternative factor affecting performance is the mixture of particle sizes coupled with the presence of natural surfactants which allow the particles to move during placing by reducing internal friction and thereby producing a modified finished product. Methods of compaction included pressing by rams, mechanical pounding and the use of hydraulically operated machinery.

The principle of the relationships of particles on the basis of size has long been understood at the larger practical level. While Roman engineering techniques in road-making aimed to produce surfaces of near flat slabs, despite the problems of jointing, 18th-century road builders had by practical experience moved to the construction point of balancing the size of components and grading them so that voids were filled and loads transmitted to material held stable by adjacent material. This, coupled with vertical grading, is the essential principle of macadam construction. Parallel practical experience convinced the builders in earth that mixtures comprised entirely of clay or of sand were unsatisfactory, whereas a balanced gradation could produce a firm and stable material. In practical terms this meant that a clay content of less than 3% or more than 20% had significant disadvantages. Too small an amount reduced the plasticity, too large an amount increased shrinkage to the point where internal fissures reduced strength. Variation in clays was found to be the most critical factor but a gradation evenly through the silts into the sands was found to be important and in many instances if the right material was not immediately available it would be tempered by the addition of sands. Further, a period of digestion in water after thorough mixing proved advantageous in terms of malleability. The occasional introduction of larger aggregate was accepted but large components were unhelpful, causing uneven shrinkage in drying. Because of varied amounts of water in different types of work this was less important in *pisé* construction and in some forms of lump than in mud brick and mortar.

It follows that in repair work earth structure is not to be thought of as a mass into which other materials can be introduced with impunity. Much of the problem lies in differential movement at constant temperature between a dimensionally stable piece of masonry and a plastic mass around it shrinking as water is removed. For this same reason large elements such as structural timber, joinery items, concrete pads or beams and other non-earth elements will induce cracking as earth structures dry. Shrinkage is also determinant of the sizes practical for the construction of brick and block, coupled with the practicalities of carrying and laying. The addition of aggregates reduces overall shrinkage proportionately. Pockets of clay between large pieces of rigid aggregate form microcracks at their interface. Tensile strength can best be provided by fibres distributed within the material and it was for this reason that chopped straw was traditionally used in the making of mud brick. A further aid to cohesion is the provision of an adequate slip plane on the underside of the mud brick, achievable by a change of material, e.g. straw

Figure 1.7 Mud Brick: The time-honoured building material now in competition with concrete block (courtesy of Dr. Paul Brown).

or a polished board or by physical loosening, where the mud brick is of the shape, size and thickness to draw itself together. The turning of mud brick during drying is related to the freeing of surfaces from the constraints of adhesion and to evening-out moisture concentration due to uneven drying. In practical terms the size limitations in making earth brick run to a maximum dimension on the flat in the order of 500 mm and a minimum thickness of 50 mm. In a brick larger than this the shrinkage is inconveniently large and in thinner material of such substantial size the tensile strength is insufficient, inducing excessive cracking. This limitation can be expressed as a ratio of 1:10 thickness to length, and applied broadly to naturally dried earth units.

While mud brick, adobe and *pisé* were all equally successful as structural methods in walling, mud brick had the special advantage when it came to vaulting in that the components could be shaped to purpose and were rigid and self-supporting when positioned. This was a crucial factor and undoubtedly caused mud brick technology construction to advance and persist. Self-supporting barrel vaults were achieved in the second millennium BC by inclined ring arch construction (pitched

vaulting). In this system the use of timber centring is avoided by construction of the first rings against (leaning back on) an end wall. Each successive ring built in advancing segments, then derives support from the previous rings. They were usually advanced segment by segment so that several rings would be actively under construction at any one time. Domed construction without centring was likewise achieved by building successive rings inclined radially to a centre – a development which seems to have followed barrel vaulting. Many earth structures carry vaults in burned brick built to similar pattern and sometimes laid in earth mortar.

Perhaps it was almost inevitable that the special demands of vaulting and arching produced the greatest specialization in shapes of mud brick. Smaller and lighter bricks tended to be produced for the construction of vaults and particularly for the building of centreless vaults using pitched vaulting. In this system, in which the successive rings were laid pitched or leaning back against an already stable support, additional adhesion would be given by producing a ribbed surface on one face of the brick by finger marking. These deep grooves increased the adhesion to the

Figure 1.8 Akda, Iran. Collapse of a vault below reveals the secondary vaulting provided to lighten the original haunching.

adjoining ring and in building, the arches were successively advanced at the lowest levels so that the vault could be built up with its highest segments supported on already stabilized lower sections. Pharaonic, Sumerian and Assyrian work exhibits this technique. Some of the largest mud bricks ever used regularly were formed as circumferential rings, each brick being a part of the segment of a circle. Ribbed vaults were built as early as 750 BC at Nus(h)-I-Jan in Iran where the entire parallel vaults of a temple were constructed with a succession of such segmental bricks meeting at a single central joint. The bricks were 1.2 m in length, some 300 mm in depth and approximately 50 mm in thickness.

Figure 1.9 Hami, Chinese Turkestan. Among the wide variety of sophisticated uses to which earth brick is put are the perforated structures used as drying chambers for sultanas. Some 25% of the wall is void to permit the flow of hot dry desert air.

Techniques of vaulting in mud brick first appeared in the third millennium BC and by the first millennium the complexities of cross-vaulting and secondary vaulting over the haunches had been well-explored and exploited. The use of very large bricks did not persist and, although there has been a long history of the formation of special types, there has been no spectacular advance in size and technique in the historical periods. The integration of fired brick in vaulting on earth brick substructures was an important part of the technology centring on Iran for a very long period. In some of the most arid parts of this region unfired mud brick components have been used with ingenuity to create grills and filigrees to fill openings, but this technique may simply be a reflection of the use of fired components for the same purpose and was effected in a rather special material incorporating lime or gypsum. Earth brick was

normally air-dried and was not normally pressed, so that the frog common to the pressing required in a fired brick is not normally found.

1.4 Mortars and blocks

Historically craftworkers have found that, as in all walling, a mortar weaker than the stones or bricks of which it is made provides the most coherent and durable structure and a mud mortar normally fills this requirement because, whatever the inherent strength of the material itself, its shrinkage on drying causes internal cracking. The mortar, being applied with a high liquid content, even perhaps beyond its liquid limit, dries by absorption of water into the adjoining bricks and shrinks accordingly. The mortar, therefore, contains multiple internal fractures which accommodate

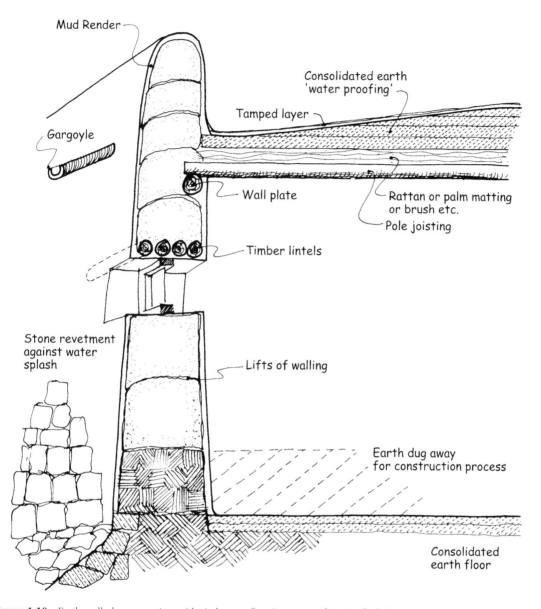

Figure 1.10 Earth-walled construction with timber roofing (courtesy of R. Deefholts).

the changing stresses upon the wall. It is therefore weaker than the mud bricks, even though it may have originated in the same mix from the same source, entirely because the separation of particles in the mortar is greater. There have, however, been many instances in which mud brick has been walled-up with setting mortars which ultimately prove more resistant to weather and survive to project from the eroded face of the wall.

Common as it is throughout the world, dried earth building is not common in Britain and in northern Europe generally, primarily because of the difficulty in manufacture. Air drying is unreliable in these climates and throughout the temperate zones of the world lump and rammed earth building are more common. Nevertheless, in both the USA and Europe there are important areas in which clay lump building has been used in temperate

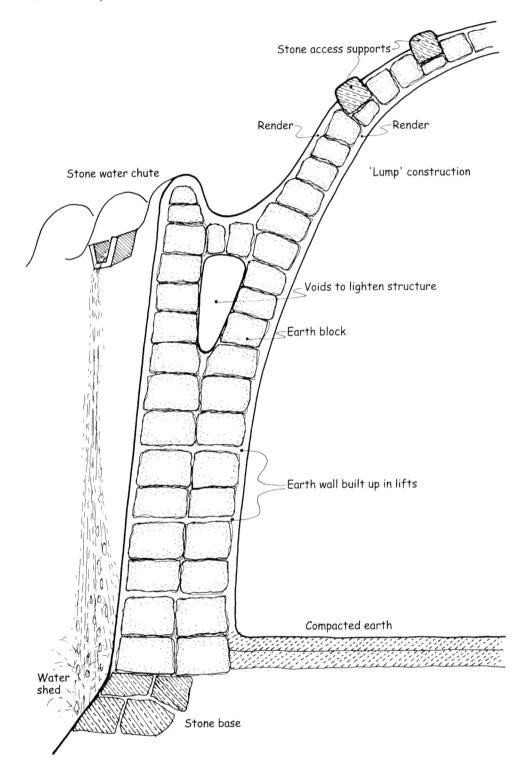

Figure 1.11 Timberless earth construction.

Figure 1.12 Samarra, Iraq. A hardened (possibly lime or gypsum-modified) mortar weathering more durably than the mud bricks bedded in it.

zones and in later centuries these have been related to techniques spread by literature or individual enthusiasms and consequently the results have been localized. The material generally has been used in as nearly dry a condition as possible and always below the shrinkage limit.

It is possible to generalize about the size of dried earth blocks despite the lack of contact between the cultures throughout the world because the general nature of the material coupled with size and weight for handling with one or two hands produces a broad optimum. A block handled with both hands will tend to be of size and weight equivalent to the modern concrete block, whereas the brick intended to be manipulated with a single hand tended to be half or a third of that size. Occasionally very large earth blocks are recorded where soils have a small clay content. Among these in Ancient Egypt are blocks as large as 960 × 30 × 300 mm. Such elements must have been treated as masonry and handled by several people simultaneously. Sled transport was common in Egypt and such an unusual and apparently inconvenient size probably related to unusual handling techniques. Far more common sizes in Ancient Egypt, as throughout the world, group into ranges in which the block would be between 300–400 mm long and 200 mm wide and the courses would be of some 100–150 mm in height. Thus blocks recorded in Ancient Egypt at Abousir are 280 × 130 × 90 mm rising to 330 × 160 × 110 mm. In other remote and unassociated areas similar sizes have been the product of the same needs. In Tibet, for instance, blocks are recorded at 350 × 180 × 150 mm, in Brazil 400 × 200 × 200 mm, and although in areas such as Antigua and Guatemala they can appear in sizes from 600 × 300 × 100 mm, they are commonly approximately 400 × 200 × 100 mm. There has never been a need for co-ordination in sizes other than on a local level and this broad conformity is a straightforward reflection of the demands of making and handling. In conservation terms replacement blocks should be manufactured to the sizes of the original work.

Despite the disadvantages of climate in the drying of earth blocks, they have been variously used in temperate zones and the early 19th century gave rise to work in North America and Europe where various relatively consistent block sizes were used. In Canada, blocks of 470 × 300 × 150 mm are recorded as normal. In the north-eastern USA (around New York) blocks of 380 × 300 × 150 mm were used. These blocks would have been twice the weight of the smaller Egyptian and Tibetan units and may well have been used in a semi-dry state similar to that of a green brick, as in some instances there is evidence of greater shrinkage away from the mortar joint than would derive from drying of the mortar alone. There are obvious disadvantages in such use but the effects of shrinkage are somewhat mitigated where the block is loaded by further work above, as the bulk of the shrinkage will then take place in a vertical direction. Clay-rich soils in a plastic condition have been used throughout history for this type of construction; the deformation of the drying clay under load produces a coherence between the lumps after placing. Not uncommonly, lumps were placed in a soft condition and shaped to encourage the deformation which would provide coherence, being formed as 'lobes' or even as 'doughnuts', and this type of construction is recorded in northern and central Europe in the 18th, 19th and 20th centuries. From written documents dating from the late 18th century it can be seen that tens of thousands of domestic buildings were constructed in this way: the technique continued and received a particular stimulus in the periods after the two World Wars. At this time it was known in Germany as the Dünner method and was used in parallel with dried block and rammed earth construction.

The use of earth building in North America is as old as the arrival and spread of Asian races southwards in the continent. Earth buildings received further impetus with the European immigrations following the Spanish and French invasions, with native Mediterranean techniques being applied in the urge to build fast and widely. Rubble in earth mortar and rammed earth was widely used in Spanish-speaking America, rendered with lime and cement mortars to create durable buildings.

One further impetus followed in the 19th century. French enthusiasm for *pisé* was rationalized and promoted in a series of publications partly through the intermediary of English agriculturalists. François Cointeraux developed the techniques he learned in his native Lyons, moving to Paris and promoting and publishing on the subject with the help of the Société Royale d'Agriculture. Before the Napoleonic wars British agriculturalists had taken up Cointeraux' ideas, and the architect Henry Holland republished edited versions of his works in the *Communications of the Board of Agriculture*. Holland had the influential support of the Duke of Bedford (of Woburn Abbey estate). In Germany the use of earth buildings in all forms was promoted in part as a consequence of the immense destruction of woodland in the industrial era. Cointeraux received a more enthusiastic reception in Saxony than in his native France.

In North America these ideas fell on fertile ground in the north-western states, the area which was most developed and most receptive to European ideas. Stephen Johnson published his own revisions of Holland's

Figure 1.13 Santo Domingo, the Convent of San Francisco (16th century). A post-medieval Caribbean example of Spanish earth construction. The limestone-in-earth plinth is topped by stabilized earth compacted in layers into tile moulds constructed with broken joints.

versions of Cointeraux' ideas, and the editor of the American magazine *Farmer*, John S. Skinner, gave regular and generous coverage to the techniques and advantages of *pisé* building, in several different forms from 1820 to 1830. Developing as they went, the new ideas spread up and down the eastern seaboard of the USA and developed a local variation, tapia, incorporating rudaceous material and limes. Many of these buildings were faced with lime-based and cementicious renders, as were their adobe contemporaries in Mexico. Numbers of them survive in the north-eastern states in the guise of rendered brick or stone.

In Africa, where mud or earth has been a universal structural material, many traditional buildings were based on wickerwork framing on to which earth was daubed. This system resulted in an entirely load-bearing earth structure in which the original timber was no more than an armature, unlike the panels of wattle and daub to be found in temperate climates.

1.5 Earth structures in Britain

Setting aside earth works (prehistoric and medieval), the wide variety of building materials available in Britain and its complex geology have produced an unusual regional complexity in earth and chalk building. Builders with direct access to plentiful supplies of stone or timber would turn to these materials, with the result that earth-building traditions in such areas were less strong or even absent, although timber framing was always accompanied by the special technique of daubing the infill panels.

Two structural techniques are generally absent from work in Britain – *pisé*, which has only been used rather self-consciously except in the East Anglian tradition, and the use of flat mud brick. Preformed earth blocks had a significant period of use in East Anglia in the 19th and early 20th centuries.

Since the advantage of soil is its immediate availability, its use was widespread but often localized and the techniques were often similarly concentrated. There are also many marginal techniques such as the use of turves and soil bonding for rubble. The principal areas in which the British soil structures are recognized are intermittently through central Ireland, the eastern lowlands of Scotland, the Solway Firth, sections of the English Midlands, East Anglia and lowland areas of the south of England stretching through to Devon and Cornwall. In all these zones the work is localized and the pattern follows the geology, but principal building types can be identified and they give rise to some quite distinctly different forms of varying wall thicknesses and forms of roof construction. A significant part of the British tradition is the use of structural timbers to deliver the roof loads either to ground level or to lower parts of the walls. This is achieved by pairs of heavy timbers forming a truss. They are part of the building traditions of the west of England from the south coast to the northernmost counties, but are rarely encountered in the east. They are known as crucks.

Crucks are regularly incorporated into the earth-building tradition of the Solway Firth where they were reared as massive frames and the external walls were quickly built around them by large labour forces gathered

especially for the purpose. To achieve this, a boulder base in an earth matrix formed the foundation for a thick wall laid in very thin layers interspersed with straw. The treading of the earth on to the wall probably forced the water out along the straw beds but also provided the tensile strength to minimize deformation during construction. Rounded pebbles and small stone aggregate were common in this material and the use of dung in the mix was common, as it provided a more plastic and workable substance than the untreated soil. The base material was non-calcareous, being a glacial drift from the erosions of granitic soils and rocks and it was, therefore, much less capable of providing a 'set' than materials with a chalk or limestone content. There is a well-known tradition throughout the country in which materials scraped from the surface of dirt roads were used for building purposes. Apart from its ready availability the road scrapings were effectively a byproduct of the road itself and its usage and repair. Soft mud squeezed up through the road was not only very well-worked and compounded with silt and aggregate fractions but it was unwanted. By removing the muddy material, perhaps with a purpose-made horse-drawn scraper, the road-maker was able to improve the proportion of aggregate in the road bed, producing a harder surface and getting rid of a surface slurry. Taking away the softer material and adding back some of the valuable hardcore would achieve the desired road finish and a surplus usable for a cottage or outbuilding. The process, perhaps, illustrates the merit of working earths thoroughly prior to placing.

On the eastern side of England, earth-building techniques made consistent use of timber reinforcement or armatures. Drift deposits sustained the soil structure traditions of northern England, north of Lancashire and through Durham and Northumberland into Scotland. These traditions have, with the exception of the areas around the Solway Firth, almost entirely vanished. Some survive north of the Humber in Holderness. Southwards they merge with the 'mud and stud' traditions of central Lincolnshire. The boulder clays found as far south as Holderness give way in central England to lias clays and through this zone are seen the northernmost outcrops of calcareous

Figure 1.14 East Anglia, UK block construction. A cottage in the abandoned village of Purton Green where the loss of decaying pink-washed render reveals the earth block (locally clay lump) construction beneath.

soils occurring as limestones with occasional ferruginous deposits. The ferrous-carbonate bound sand-based soils to be found in northern Oxfordshire are unusual even in this area and contain minimal clays. However, they benefit from setting qualities which have made them surprisingly durable despite the high silt/sand content. Keuper marl, found across northern Leicestershire, has provided the only known surviving Roman mud brick in Britain. Soils of this type were used as far west as

Warwickshire and as far east as Rutland. Local names for these types of construction have proliferated and many have largely or completely died out, but in the south, three building types are still well-known and identified by local terms – clay lump in East Anglia, wychert in Buckinghamshire running through across the Thames valley to Wiltshire, and cob in south-west England.

Clay lump is the only significant tradition of building with clay brick or block in Britain:

the Roman technique, based on Mediterranean practice failed to survive the demise of the Empire. Lump construction stands apart from other earth-building methods, providing relatively thin vertical faced walls well-aligned mechanically and frequently dressed with fired bricks at quoins and openings. Sometimes it was faced in brick and unrecognized examples are still being discovered. It was a late technique, having arrived in the 19th century, when it was in competition with the increasingly mechanized brick-making industry using the readily available coal supplies from industrial England. The clay lump tradition merged with and overtook an older local tradition of compacted mud walling. It made use of the patchy outcrops of calcareous soils resulting from the glacial retreats and erosion of limestone overburdens. It was generally limited to East Anglia.

In central southern England, Buckinghamshire, Oxfordshire and across the downlands of Wiltshire, the longer-known techniques of mud walling persisted throughout the 19th century. Localized deposits of clays in hollows in chalk landscape would contain high proportions of calcareous material and, suitably pounded and prepared, they had the advantage of achieving a rigid form quite rapidly after placing. They also provided a particularly good bond to limewashed surfacing and to limebased renders. These ideas strayed into the downland areas of the south-east, often incorporating nodular flint, particularly where the materials were used for estate or garden walling.

Chalk-rich earths fall into a special category. The soft rock, formed from the compacted marine deposits of sea creatures, has been eroded and deposited again with topsoils as a mixture of pulverized chalk, sand, silts and perhaps some clay. Compacted as walling it is one of the most rigid and durable types of earth structure known – a phenomenon which appears to be due in large part to the interacting angular shapes of the foramanifera whose shelly surfaces interlock, coupled with the redeposition of dissolved calcium carbonate which may lock them into a crystalline matrix or provide crystal growths around their points of interaction, which effectively hold them in position.

The deposition of the chalk deposits arises from minute and barely visible secretions from algae (coccoliths) which form the mass into which shells from very small shellfish (foramanifera) have, over an immense period of time, dropped to form a mass of nearly pure calcium carbonate. Silica may cohere in large nodules within the chalk as flint, and iron sulphides as pyrites. Where sand has been deposited evenly, hard chalks may result, providing a moderately useful building stone.

Westwards, where the chalk and limestone gave way to the Devonian shales and culm measures, a rich and persistent tradition of heavy walling known as cob has survived long enough to remain a living skill. Whereas the chalk and clay mixes of the downlands are extremely sticky and set hard when drying, the silty Devonian soils compact well, are more easily handled and often contain flat fragments of shale, known to its users as skillet. By virtue of the techniques of laying, these components probably stratify and provide some rigidity across a wall structure. Cob building can be carried out with quite substantial lifts, on occasion sometimes as much as 1 metre or more, the material being unrestrained by any form of shuttering. The common technique involves building to a greater width than required, the face being trimmed back to the alignment sought, often solely by eye.

The one other traditional English usage of clay earths in building is daub used in panels that provided both internal and external walls in timber-framed buildings. While not confined to the eastern and southern counties, it was a tradition which necessarily spanned many geological zones and because the volumes of material required were less and the performance requirements were greater, the use of the material was not as geologically precise as with mass earth walling. Clays could be carried to areas of sandy soil to provide the right mix for a daub and sand would be introduced, sometimes from a considerable distance, to ameliorate the more recalcitrant clays. There were many mixes for daub but, unlike most mass earth constructions, they almost invariably included straw and sometimes other fibre. It was never satisfactory to have high shrinkage with the resultant cracks around the edges of panels, so sand/silts or even loams were used to modify clay-rich mixtures, while lime or crushed chalk

Figure 1.15 Dunsters Mill House, Sussex. Medieval wattle and daub panelling fitted into the complex shapes in spandrel and panel within oak framing.

was often also deliberately added to provide greater stability and strength.

When building with earth block was introduced into England early in the 19th century, it gained popularity in some areas for low-cost, low-rise construction, although on occasion it was even used for the towers of windmills. Generally it was known as clay lump. Its use was largely confined to southern Norfolk, northern Suffolk, and part of Cambridgeshire and these quite distinct boundaries as to time and area derive from the availability of a suitable earth mixture. More particularly its use relates to the stimulus given by publications relating to agricultural improvements and to the exchange of knowledge among agricultural 'improvers'. Blocks became standardized and were compatible with fired brick which enabled the materials to be used together; sometimes the bricks provided as a complete skin and elsewhere as quoins, foundations or arches.

Figure 1.16 Parthings – a house in West Sussex, with Medieval close-studding where the infill panel of wattle and daub is reduced to a single stave.

Blocks 18" (460 mm) × 8½" (215 mm) × 6" (150 mm) were the most common. Although many early smoke hoods and upper chimneys were built in wattle and daub, fireplaces were always built in brick, and brick was normally used as a plinth and foundation. Both lime mortar and earth mortars were used and, provided that the block was firm and hard, it was not necessary that it was dry. A higher moisture content could be tolerated in a block being used in a lime mortar. Where an earth

mortar was used it was convenient that the block was drier than the mortar so that water absorption by the block would stabilize the bedding and prevent it being squeezed out. The reduction of water content in a lime mortar was not detrimental in the long term as carbonation continued in normal atmospheric conditions. Walls were always protected, usually by a render of lime and sand, although occasionally a tar finish was substituted or applied over a render coat to

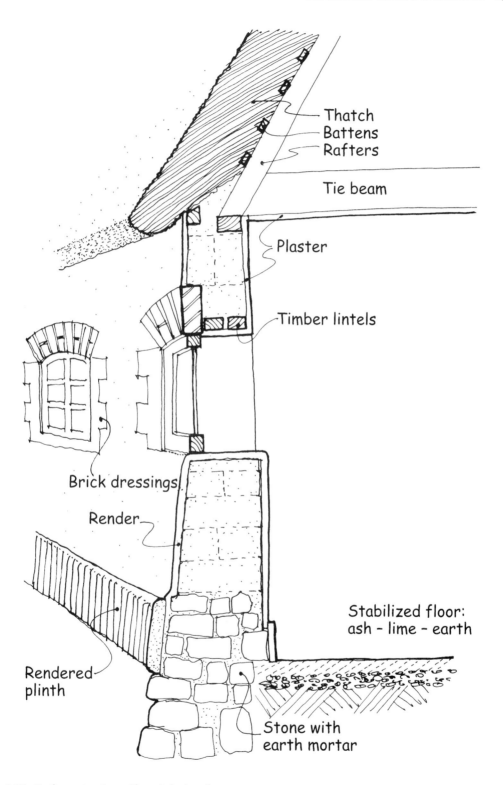

Thatch
Battens
Rafters

Tie beam

Plaster

Timber lintels

Brick dressings

Render

Stabilized floor:
ash – lime – earth

Rendered
plinth

Stone with
earth mortar

Figure 1.17   Earth construction with a pitched roof.

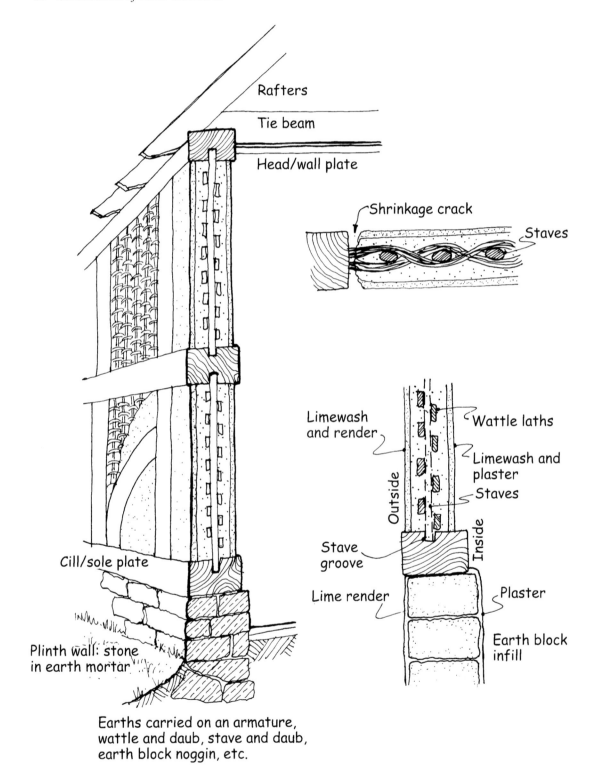

Rafters

Tie beam

Head/wall plate

Shrinkage crack

Staves

Limewash
and render

Wattle laths

Limewash and
plaster

Staves

Outside

Inside

Stave
groove

Cill/sole plate

Lime render

Plaster

Earth block
infill

Plinth wall: stone
in earth mortar

Earths carried on an armature,
wattle and daub, stave and daub,
earth block noggin, etc.

Figure 1.18 Framed earth construction.

provide adequate adhesion. Keys were sometimes formed in the face of the block by applying lengths of timber to the inner face of the mould. These buildings were traditionally limewashed and lime wash was occasionally applied over a tar finish with a sand key. The washes were coloured on occasion and quite deep ochre and brick-red tones have been recorded in this material, perhaps in an effort to emulate brick building.

The more basic technique of piling moist earth on to a wall, treading it down and trimming the face has been used in Britain from time immemorial but the surviving buildings in this technique used world-wide are predominantly to be found in the south-west of the British Isles. Their importance and usefulness are now recognized and their conservation is generally assured. English building technology in lump and cob construction was essentially a part of the north European tradition and has been very basic and straightforward, as befits the vernacular simplicity of the buildings for which it was employed. Openings were generally spanned by timber lintels and wall thicknesses were robust rather than daring, but this is by no means true of the equivalent construction elsewhere, particularly in the Middle East. The one fundamental and prime advantage of the mud brick or adobe was its rigidity at the time of placement. On this factor depended the ability to build the vaults, arches and domes which turned earth into a structural roof.

Two of the most characteristic forms, both found in England, are wattle and daub, and mud and stud. Common usage of the latter construction is confined to the eastern counties – Lincolnshire in particular – and produces a relatively heavy wall whose core is a series of timber members joining rails above and below. In Durham and Northumberland the mud lumps placed in plastic condition between the staves were known as *cats*. Wattle and daub construction is a common north European technique ubiquitous in medieval northern Europe where timber framing was the ready and most effective means of providing dwelling space. Commonly it was formed by fixing vertical members between the horizontal structural rails in a panelled wall. These were set at a convenient distance to allow more pliable members to be woven between them. The

Figure 1.19 Solid earth construction. Section of wall where the outer skin remains undamaged at the upper level, demonstrating the successive application of layers of plastic earth forked on to the wall (courtesy of J.R. Harrison, architect).

technique was analogous to forming a cloth in which the vertical members were the equivalent of the warp and the more pliable members formed the weft. Nails were not required; the staves were fixed by placing them in holes bored in the upper rail and driving them tight along a groove in the lower. Cleft oak was the best material for this work, the heartwood being particularly resistant to insect attack, but in many instances the more pliable members were made of hedgerow material such as hazel which succumbed much more readily to the beetle and was often ineffective after a generation or two. The daub

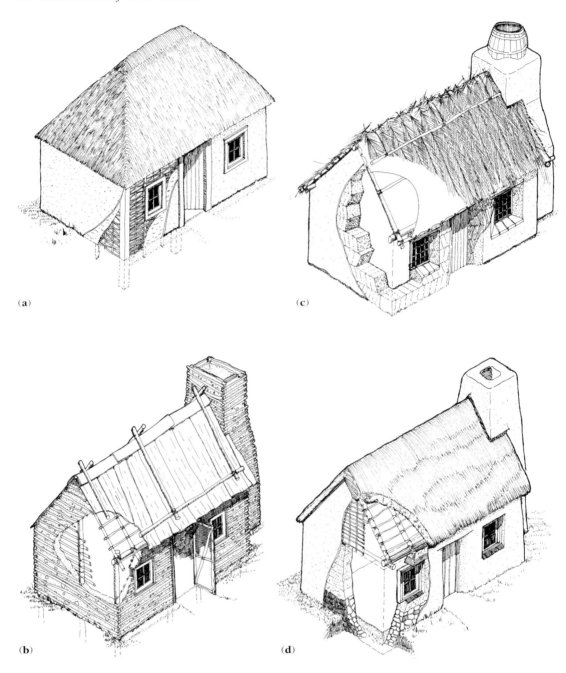

Figure 1.20 a–f British earth buildings in microcosm (from Jeremy Salmond's *Old New Zealand Houses*).
(**a**)   Post-fast frames with wattle and daub infill. Wattle and daub is typical of the south and east of England.
(**b**)   Post-fast frame infilled with mud and stud construction, found extensively in Lincolnshire. Note: the term *mud and stud*, as recently applied in the UK, tends to be specifically used for construction in which the studs are closely spaced, forming a palisade with or without laths on to which the mud is applied.
(**c**)   Sod construction, extensively used in Scotland and Ireland.
(**d**)   Cob construction – predominantly a west country technique.

(e)

(f)

(**e**)   Mud and stud house at Robin Hood Bay, Marlborough.
(**f**)   Corner detail.

Figure 1.21 Quedlinburg, ninth-century capital of Saxony. The extant timber-framed town displays the several types of wood-reinforced earth walling traditional across northern Europe.

Figure 1.22 Quedlinburg, 16th century. Conventional wattle and daub above a panel of earth brick noggin in timber framing.

would be applied simultaneously from both sides to minimize vibration. Sometimes it was thrown with considerable force and its plasticity was enhanced by the use of dung or some other albuminous material to provide a glutinous additive. Subsequent working would both smooth the material and force it into all the crevices in the wattle panel, but by virtue of the elasticity of the panel the daub would always be left finally with an undulating or uneven surface which became characteristic of the material.

In France a 'ladder' of horizontal supports between vertical studs has been the method of securing earth walling in villages of Champagne and Alsace. Elsewhere these ladders have become vestigial, providing simply a series of pegs projecting from each side of every stud, with entirely satisfactory results where the compacted earth is suffi-

ciently thick. Throughout the world, variety and practical ingenuity have been the prime characteristics of this combination of materials.

1.6 Construction methods

The simple shell of a paraboloid dome has been refined in mud brick construction to yield slender membranes used widely across Africa and in remarkably refined forms in northern Syria (see Fig. 1.9). These shells, built without centring are the product of many generations of evolution. They are continually rendered with a mud–dung plaster to keep them in repair and are provided with projecting stone supports for maintenance purposes. Each follows its own geometry. Sub-Saharan Africa contains many similar localized techniques.

Figure 1.23 Quedlinburg, 17th century. A vertical weave across horizontal staves in timber framing with external and internal earth render painted.

Figure 1.24 Quedlinburg, probably late 17th century. Reed lathing on timber-framing carrying lime stabilized earth render externally and internally.

A very much more sophisticated structural philosophy has been developed over the centuries on the Iranian plateau and in immediately associated areas where the complexities of domes set upon pendentives of ribbed vaulting and interlocking arching have demanded accuracy of construction and a high-level of skill in execution. These forms of construction, employing a wide range of specially made shapes of brick and dependent upon ribs for the support of intervening membranes, stand at the heights of technology in earth building.

The density of the earth fundamentally affects its strength and has some effect upon its behaviour and weathering. The density essentially depends upon the degree of compaction at the time of making the material and/or placing. Minor variations of density flow from the type of material, its curing and methods of mixture, but the significant differ-ences derive from physical compression. Most block and lump constructions have throughout history been made without significant compression; the prepared material has been placed in moulds or forms and subject only to modest pressures from hand-held tools, or at most by trampling.

There are two methods by which clays and earths are normally consolidated during manufacture and placing – mechanical (usually hydraulic) and physical impact (usually pounding or ramming). Mechanical methods of compaction for earth structures can be disregarded in historical and therefore in conservation terms. The pressed earth block may be regarded as a modern material and the densities of earth lump, daub and plaster, earth mortar and mud brick may be bracketed together. The dry density of earths normally used in building lies in the range 2.1–2.3.

Figure 1.25 Faraj, Iran. Arch construction in mud brick in the domed substructure of a destroyed palace. Gypsum-stabilized bricks define the structural alignments.

Precise figures become meaningful when the porosity, content, volume and density of any large aggregate are known. The true density of the aggregate has a bearing on the density of blocks (limestone, for instance, is less dense than granite) but the nature of the natural compaction is more important in overall calculations. The pounding or ramming of earth placed into a structure may be regarded as an extension of the basic earth-building technique which simply involved placing the prepared material on the wall in a semi-plastic state, treading or pressing it into position so that it melds with the already placed material and trimming off to provide the exact shape required. It is natural in this process for builders working on a wall, having had the material thrown up to them, to place it with a tool and tread it in with their feet to press it into position. However, this can only be done to a limited extent without impairing the stability of the construction, but the advantages of

Figure 1.26 Faraj, Iran. A succession of broken vaults illustrates the effective ground coverage of thin-shelled vault construction.

greater compaction will immediately be evident. The natural immediate response is to place timber shuttering either side of the wall in sufficient lengths to allow the shutters to be connected so that pressures upon them are neutralized by tension in the connectors. This simple and inevitable development has made pounded earth a form of construction encountered worldwide and its users relatively quickly discovered that a stiffer or less wet mixture could be employed with the immediate advantage that shrinkage was reduced. This was concomitant with the secondary advantage of greater strength on hardening.

Since clay–aggregate mixtures may be thixotropic, i.e. flow under vibration and become more rigid as vibration ceases, the effects of pounding can be that:

- gases (air) included in the material were expelled usually as small bubbles rising to the top

- the particles moved into closer association and aligned themselves more logically
- the particle alignment encouraged crystalline development in the clays
- the greater density produced higher compressive and tensile strength
- the cohesion between material placed at different stages would be virtually total as opposed to the substantially weaker bond produced where a mortar was employed

There can be dangers in the method. When a mix becomes fluid by vibration, forces are transmitted in all directions because the fluid is the hydraulic medium. Army engineers in charge of enthusiastic platoons had to warn them that pounding must not be rhythmical or co-ordinated lest the shutters burst, and excessive pounding, like excessive vibration in concrete, would tend to segregate the material, particularly if there was a significant moisture content.

Figure 1.27 Layered earth construction in a modern building in the Yemen. The structural lifts are clearly demonstrated in the finish, where the fibre content is drawn down in preparation for the succeeding lift (courtesy of Dr Paul Brown).

Figure 1.28 *Pisé* in the ancient Far East.

The greater the level of organization and scientific thought, the more likely that a compressive technique would be employed. The resulting material is frequently known as *pisé*, as a result of its extensive use by French engineers. In its simplest form the technique has been used in the Far East for two millennia. Permanent shuttering of fired brick might be employed; the earth fill is an economy in time, fuel and labour. Much of China's Great Wall was built in this way. The needs of colonialists and the military for the maximum construction with the minimum investment led to its extensive and widespread use in the 19th century, with revivals under economic pressures and following the enthusiasm of surveyors, architects and engineers in the 20th century. But the technique was by no means confined to the literate, the disciplined and the organized. It is found among vernacular builders throughout the world and with its greater load-bearing capacity has offered the

Figure 1.29 Bosnia. Typically Balkan timber-framed construction in which clay block forms the noggin. In composite structures of this type the structural rigidity of the frame can be significantly enhanced by the rigidity of the panels which can be load-bearing in both lateral and vertical directions.

opportunity to build high and impressive structures in the widely differing conditions of remote wadis of the Arabian peninsula and the high Himalayas. The limitation of the technique is immediately apparent: it is effective as walling and in massive structures such as dams, embankments, revetments and retaining walls. It cannot be used for thin shell structures or ribs.

Internationally, the large bulk of earth building has been by the simplest and the most basic method, piling on to a wall a mixture of clayey earth and sands puddled with straw and made sufficiently soft and plastic with water by working. There is little complication in technique and the tools were straightforward implements of agricultural type – forks, spades, shovels and iron blades on handles similar to hoes. Some special cutting knives were produced for the purpose of paring down the surfaces. These techniques of hand-working gave builders the opportunity to build the most fluid and plastic-shaped walls. The resulting curves were further accentuated by the renders and mortars used to provide the surfaces with a weathering. It is in these soft and subtle contours that much of the distinctive charm of earth building can lie.

One final range of structural types exhibits perhaps the widest variation of earth building, being dependent upon a framework or armature for its shape and structural rigidity. All earth-building techniques have an immediacy and essential simplicity which is one of their most appealing characteristics. The use of armatures has effectively been a consequence of the availability and usefulness of earths in their plastic state to seal the voids, surfaces and junctions in lattices or frameworks. Like sprayed concrete, the earth becomes a structure in its own right once it is sufficiently thick, irrespective of the armature.

The builders of the reed houses who today coat their high barrel-vaulted wicker halls with mud are following a regional tradition of four or more millennia. Over wide areas of Europe and Asia, structures of timber post and rail, some filled with wattle and others with parallel vertical strutting in a variety of forms, have been used as the basis for the support of a skin of plaster daub of similar composition to the materials used for walling, whose purpose has essentially been that of weather-sealing the masses of applied mud or earth which have themselves become structural. Timber decayed or attacked by insects has powdered into uselessness, leaving the earth panel hollow to the core but sufficiently coherent to remain functional. Much earth daubing, however, is set on panels which are held in position by timber framing.

Well-applied and properly maintained panels of earth-based infilling can have an indefinite life subject to the survival of the framing which holds them in place. They can rarely be structural in the sense of being load-bearing but this is not their function. They are often the most easily damaged part of the fabric but their conservation is important in the broader picture because very often they present surfaces that give special characteristics to the building itself.

A short-term view of building construction may justify the use of earth, brick, timber and mortar in close combination. The pioneers who built their stockades with timber driven into the hurriedly raised earth banks were entirely justified in using such materials by the short expectancy of its life. Those who want to conserve the forts they produced have to make do without the perished timber, now vanished, or risk the censure of that part of the conservation movement which regards the reassembly of surviving components (anastylosis) as legitimate while disapproving of the replacement of known but vanished structural elements. The intimate mixture of timber with porous masonry produces many a problem of this sort. Of all the combinations of organic with inorganic materials, the presence of timber produces one of the greatest ethical and practical problems in conservation. The fundamentally practical aspect is simply the long-term incompatibility of these materials due to differing rates of decay.

While for practical purposes much earth construction may be regarded as having an indefinite life, this is much less true of timber and the rates of decay which may be anticipated of the two groups of materials diverge appreciably over a long time span. A strong case can be made for the removal of timber from an intimate relationship with moisture-bearing earths where its failure can be anticipated, but there are many instances where the relationship is so interdependent that it must be maintained. Thus, while it is practical and feasible to replace with solid brickwork a piece of timber built into a substantial brick wall, it is much less possible to replace the wattle framework of a wattle and daub panel without destroying the panel in its entirety, thereby destroying the historic quality of the surfaces and structure. Elsewhere the timber may be an essential visual or structural component which cannot be discarded without historical loss.

In exploring these relationships and in making conservation judgements a balanced perspective and discretion are the essential guide.

1.7 The structure of building earths

Discounting those earths which contain bituminous materials, there are effectively two types of mix which have over the ages proved suitable for building work – those containing calcite (chalk, lime or limestone) and those without. In the first, crystalline formations provide a mechanism for solidification and coherence and the group could be broadened to include gypsum-based materials where these occur. The mechanism involves the inclusion of available water, which enters the compound at molecular level as water of crystallization and becomes bound up in the material itself. Where calcite is not present, although some crystallizsations can occur, friction is the predominant mechanism of coherence and the tensile strengths of these materials are, therefore, generally lower.

It is common practical experience that, where friction is the prime method of transfer of stress, variation in size produces the most stable results. This is true of material compacted into foundations and road bases and on the very much smaller scale it is true of earths. The mechanics of this phenomenon depend upon a triangular distribution of forces repeated many times which produces an inter-distribution of forces. Technically this is known as *close packing*.

1.8 Methodology

Most conservators work by the instinctive application of technique, that is to say, the combination of experience and knowledge which allows them to apply a methodology to their actions without consciously analysing the sequence or the reasons. There is, however, a very strong discipline and logic in the processing of repair work of historic buildings which can be defined as a method. As applied to earths, the nature and the sequence of decision-making are correlated in the following way:

1. Repair must be justifiable as being both necessary and the minimum required to achieve the sustainable result, while at the same time meeting criteria of reversibility and integrity. Qualities of sympathy, compatibility and the manifold practical aspects play a part in the decision-making, while more subtle factors such as local employment, maintenance of tradition, training and the coherence of grouped buildings must be significant.
2. The structure must be clearly understood in its historical background, the forces acting upon it and the nature of construction. This may involve testing to determine the soil content, clay type, pH value, strength of materials, climate, usage as well as the analyses which go with historical evaluation. The building must be appropriately recorded so that changes can be identified and the reasons for failure must be investigated fundamentally so that the cause can be appropriately dealt with in conjunction with making good loss or damage.
3. The repair or intervention must take account of the causes of damage and be designed to resist recurring failure, and to make good effectively. It is important to achieve a balance between the materials required for the intervention in practical terms and the requirements of tradition. There is, for instance, no point in using a matching clay to make good a repair at the base of a wall if that clay is expansive and will shrink away in the process of drying so that it offers no structural support. Better to use a compatibly soft, inert material which will provide the support required and be durable. In the opposite sense, however, there is equally no point in buying an expensive resin as a consolidant to an earth wall whose surface is simply an eroded core which has lost its protective render. No historic surface is being protected and it is better to return to the native technique.
4. Temporary support required to achieve the repair and ensure the stability of the building must be allowed for at the outset, together with the protection of the structure during the repair period.
5. In managing the project it is important to ensure that all approvals have been obtained, that materials and labour are available and that the work has been properly costed and supervised.
6. The long-term effects of any intervention and decay of materials must be considered and applied to a programme of maintenance.
7. Finally the work must be documented and archived.

Structural failures in earth buildings usually fall into four categories:

• undercutting due to erosion and/or salt decay
• failure of timber components or loss of bearing by harder material, such as arches, lintels and quoins
• vertical cracking due to differential movement or external forces, perhaps accentuated by water
• ingress of water

Secondary failures of vaults, floors and other structures may follow. A general principle in supporting earth structures must be that the restraint is delivered over as wide an area as possible and with a uniformity which is usually best achieved by the use of softings – foam pads, textiles, sand bags, flexible timber and resilient sheeting. The essential principle is that no point load is delivered which approaches the maximum compression strength of the earth structure. If this principle is not observed the point load will punch through the earth. Temporary support must, therefore, always be designed on the strutting and planking principle that the restraining force delivered through a rod or pole is spread on to the wall by a plank or sheet and it may well be necessary to interpose between this spreader and a wall, arch or other structure soft packing materials, which can now include sheets of plastic foam or injected expanding foams, later to be removed. Arches must always be supported uniformly and never by a localized thrust. Although mud structures are relatively easily cut through and where they do not contain stones can be bored, needling is not easy and will be used as a last rather than a first resort.

The flexibility of earth structures permits them to be manipulated more readily than more rigid forms of construction other than framed buildings. Walls leaning as the result of basal failures may be restored by taking shores over an extended length, jacked forward gently and evenly to permit reconstruction of the lower zones which have been undercut or are beginning to fail, perhaps through the effects of ground water or deposited salts. Stabilization can then proceed by reconstruction and insertion.

1.9 Available options

The first option considered will normally be a traditional form of repair, that is to say, the replacement of lost structural earths or renders with like material. While traditional rerendering is straightforward and satisfactory in principle and in effect – other than the extensive labour involved – traditional methods of structural repair by infilling are frequently unsuccessful because the introduction of plastic earths into voids is beset by the problem of the shrinkage of the inserted material and bonding failures in consequence. It is for this reason that many native techniques of repair involve the insertion of stones or even boulders and the modern equivalent is usually concrete block and/or liquid concrete which

likewise is incompatible with earths and frequently leads to failure.

Where gap filling is necessary careful design can overcome the problem by selecting alternative fillers which do not shrink, by minimizing water in the insertion so that the material is below the plastic limit, by improved bonding techniques and by the use of synthetic materials.

Non-contracting fillers may be designed using clays with minimal expansion – the kaolinite ranges – by the addition of limes and gypsum whose expansion can counter contractions and by the use of inert materials such as brick dust and pulverized fuel ash. Introduced material may be tamped into small cracks, a grout may be run into voids or it may be introduced in a stiff plastic form into larger openings. By careful design of materials, varying degrees of hardness, solidity and setting may be achieved with appropriately small expansion or contraction characteristics.

By the inclusion of minimal water, contraction can be reduced or controlled. Where the material can be adequately contained dry packing may be used, the appropriate mixture being fed into voids as a dry powder, tamped and then allowed to take up moisture without flooding. This may be achieved by sustaining moist external conditions for a period using absorbent superficial material such as sacking held in position against the structural surface by temporary supports. A coherent fill can be attained by methods of this sort, although high strengths should not be anticipated.

Introducing material in a plastic condition leads to the problem of workability which demands a water content below the plastic limit. A relatively dry material is difficult to introduce into small voids: a wet material will show increased shrinkage. In addition, a dry wall structure will fail to bond to inserted material whereas the introduction of water into the existing earths will exacerbate the shrinkage problem and leave lines of weakness at the junction of the fill.

Where only one surface requires bonding the problem is relatively straightforward and wetting the existing structure together with the introduction of a suitably plastic mix can be expected to work if the intervention is small. Where substantial voids have to be filled and bonding is required on two or three sides the

Figure 1.30 a–b Two examples of typical surfaces of mud render, one weathered and the other newly applied, showing the patterns of inherent cracking which become the foci of erosion.

most respected method is to form the inserted material into an appropriately shaped block which is then placed dry with an appropriate mortar of earth or a similar weak mix being used to pack around and behind and provide necessary bedding.

Bonding is sometimes advocated. This may be achieved physically by introducing bars driven into the existing structure, around which the new material is laid and packed or even, as has sometimes been advocated, by the introduction of tile coursing or creasing. Physical bonding of this type has the advantage that the keying is mechanical and continuity is maintained even if an adhesive bonding is not sustained. Tiles, rods, metal connectors and similar devices are not naturally compatible with earths and it must be accepted that any movements transmitted to and through them will not readily be trans-

ferred to the main wall structure, but will tend to be expressed as a movement between the harder material and the softer.

There is a place in these considerations for synthetic fillers. Joints which are the result of movement between sections of buildings and joints where an internal pressure is desirable or necessary may be best dealt with using injected foam techniques. Proprietary techniques are available by which reagents and resins are mixed at the point of application to provide foaming setting compounds such as polyurethanes. These may be formulated in varying degrees of elasticity. Their disadvantage is impermanence. They should always be protected from ultraviolet light and even in protected positions they can fail due to chemical instability over a period of time. They cannot, therefore, be relied upon for structural continuity in the longer term but they can provide a basis for expansion joint-ing. Circumstances in which rigid materials are an inherent part of construction, as in the case of brick and stucco facings, may provide circumstances in which synthetic fillers and synthetic bonding agents may be alternatives to be considered.

1.10 Methods of workmanship

Traditional methods of workmanship have the merit of being long tried and in many cases established as the most effective method of using traditional materials. Modern equipment may provide short cuts or labour-saving amendments to traditional techniques – the mechanical chopping of straw or synthetic fibres, mechanical methods of mixing and mechanical haulage, placing and pumping may be combined with modern methods of shuttering, lifting, compression, scaffolding and so on. Such methods which are simply variants on established technology may be used with predictable effect and success whereas other techniques of more recent invention, such as high-frequency mechanical or even ultrasonic vibration may have appli-cations in special circumstances yet to be proven.

In traditional techniques considerable force is often applied in the delivery of plastic mater-ial. Wattle and daub panels may be effectively coated using the simultaneous application of thrown daub from either side and the rhyth-mic placing or pounding of mass earths where complete sections of structure are to be renewed may be applicable in repair on a large scale. Where there is any doubt about the effects of vibration and the necessary force required to introduce the repair materials, adequate shuttering should always be consid-ered, the surfaces of all the restraining materi-als being given soft packing to distribute the pressures across the wall surface, uniformly across the soffit of the vault or the floor.

In preparation of putty-like materials it is a general truth that the longer they are left in the preparatory wet condition the more pliable, workable and, therefore, ultimately more satisfactory they will be. Many such materials benefit from repeated mixing which leads to the integration of microscopic parti-

Figure 1.31 . . .Ease of construction. . . Cutting earth brick to size with a heavy knife.

cles. Adequate time should, therefore, always be specified in the preparation of materials of this type. With the introduction of reinforce-ment, whether it is synthetic or natural, very thorough mixing is desirable if uniformity of material is to be achieved. This is not always necessary. Some traditional earth structures are notable for their inclusion of reinforcement by layering rather than uniform mixing.

In the working of earthenwares, bonding is often achieved by wetting – as in the forma-tion of pottery – and the same may be true for some earth structures, particularly in the adhesion of rendered coatings. Since the coating will by virtue of its thickness be weaker than the wall structure to which it is applied it must move with the wall or, if partic-ularly hard, it must be held in a flexible relationship. Renders are therefore naturally and traditionally designed to accommodate wall movements by multiple fracturing within their own texture. A weak material was, therefore, commonly used and a strong bond was achieved by providing sufficient wetness on the substrate and in the render to meld the two materials. Traditional workmanship, therefore, favours adhesion achieved by wet pressure bonding – trowelling, pounding, hand-working and smoothing techniques are common. The multiple superficial cracking which follows the drying process provides a key to surface finishes such as lime-based washes or even tar, traditionally applied hot, but now available as a cold emulsion. Even as an emulsion, however, the large molecular structure of bitumens inhibits integration of the materials.

A basic knowledge of the chemical and physical characteristics of materials is important to the conservator in understanding their behaviour.

2

Constituent materials

2.1 Essential physical background

Particle sizes are described in millimetres and micrometres or microns. The term *micrometre* is tending to supersede micron as being more descriptive of a physical dimension. The symbol is 1 μm. Table 2.1 gives typical sizes which assist in understanding the scale on which phenomena and reactions take place.

Table 2.1

micron = micrometre = one millionth of a metre = 1 μm

Atoms and molecules tend to occupy spaces of 'diameter' 0.0005 μm. Particles in colloids occupy spaces between 0.001 and 1 μm. Particles in emulsions occupy spaces from 1 to 10 μm. Particles in clays are generally in the range 0.1–2 μm but may be as large as 20 μm. Particles in silts occupy the range between 20 and 6000 μm approximately (0.002–0.06 mm). Particles in sand are upwards of this size. Building sands will not contain much material below 0.5 mm or above 2 mm. Grits range from 2 to 5 mm and gravels are generally larger than 5 mm.

When placed in position, solidified earths are subject to the same processes of strain, fracture, abrasion, erosion and dissolution as created their constituent particles but they are important to humans in totally different terms, representing created artefacts which encapsulate our history and so demand repair and retention.

These particles vary in size and in nature. The general classifications in regular use are gravels, sands (including coarse sand), silts and clays. Geologists used the terms rudaceous, arenaceous and argillaceous for the same definitions and particles can generally be classified by size by thinking of them in these broad groups.

The definition of particle sizes is now a matter of general agreement with minor variations. A convenient definition is as above and may be summarized as follows:

- clay particle sizes are below 20 μm (0.002 mm)
- silts may be up to 6000 μm (0.6 mm)
- sands and grits are above this and become gravels above 5 mm

Soils with a high sand and gravel content are normally known as sands, despite a small content of silts and clays. Loams contain a higher proportion of silts and clays, and soils containing upwards of one-third clay are generally referred to as clays despite the preponderant content of fractions containing larger particles. The tendency of clays to disperse makes a soil deposit containing over 50% of a clay relatively rare. However, potters' clays and clay deposits do contain a high proportion of clay minerals and much of their non-clay mineral inclusions approach clay size, i.e. up to 0.00001 mm. Clays are a specific group geologically comprising small flat crystal structures often described as plates, with a tendency to 'stack'.

For earth construction a clay-rich material will rarely contain more than 25% clays while the leanest will be in the order of 5% in other than calcite-rich material. Any clay content richer than 25% becomes excessively plastic and pliable with too high a rate of expansion as water is absorbed. These percentages are

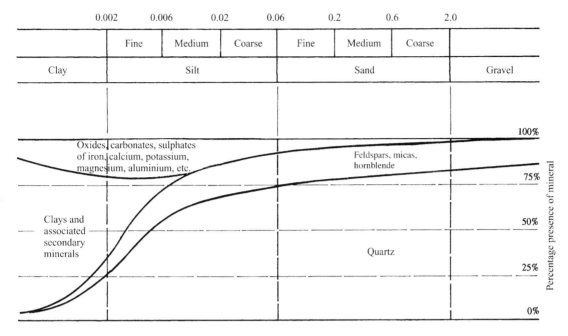

Definitions are approximate

Figure 2.1 Soil composition.

given by dry weight. A soil suitable for building in its plastic and workable state will
contain about 15–20% moisture by weight. A
dry sample of a typical construction earth with
an average clay content can take up about the
same amount of water by weight as it contains
clay while remaining firm and stable. Beyond
this point it tends towards plasticity and when
the water content has been doubled it loses
its integrity and can collapse.

2.2 The basic material

The suitability in composition of earths is not
determined simply by the volumes of clay, silt
and sand. Sands and silts themselves vary in
characteristics, the particles being sharp or
rounded, and the clays differ widely in their
behaviour. The proportions of the different
types determine the behavioural characteristics. A number of simple tests determine the
suitability of the mixtures. When the clay is in
a plastic and workable state a small roll of the
material may be made to pencil thickness. If
a length of some 50 × 3 mm is bent while

holding the ends it will break at between 45
and 90° of bend. An earlier break will indicate
an inadequate clay content and a later break
an excess.

With practice a suitable mixture can be
determined by feel: the material can be rubbed
between the fingers in a wetted condition
revealing the slipperiness of the clays and the
grittiness of sands. A sample shaken up in
excess water in a glass cylinder settles to
reveal the proportions of each component.
These and similar such field tests are sufficient
for the experienced conservator to determine
compatibility of mixes and their basic suitability for particular purposes.

Laboratory analyses used to supplement
field tests will enable the original material and
replacement or modified materials to be
assessed as to porosity, particle size, clay
chemistry, calcitic content, salinity and salt
type. Extensive programmes and the use of
unproven additives and techniques may
demand sophisticated analyses and accelerated
ageing or weathering tests, as well as field
trials. Such complex investigations will not be
the norm for the practising conservator whose

work will be governed by general experience and knowledge of the key factors in earth behaviour, such as particle size and the nature of the common mixtures of clays.

Suitable particle size distribution has been found to be in the range of one-third to two-thirds by weight of sand and gravel, one-quarter silt and one-fifth to one-tenth clays, but these proportions vary widely in practice with a calcite content sometimes contributing up to a quarter of the total, making the material more a chalk or limestone than clay. The clay fraction will vary in character with location and distribution – tropical soils produce much higher proportions of the most severely degraded clays, the laterites. These are strictly hydrated aluminas almost invariably rich in hydrated iron oxides and are not clays when analysed. However, their behaviour in a practical sense puts them into the same general category of material. Different climates produce different types of clay fractions even from similarly related original rocks. The laterites will normally be red and brown due to their iron (haematite) content, or yellowish. These clays are less responsive than others to expansion and contraction on the absorption of moisture and make a durable building material. Clays common in semi-tropical zones and as a product of the breakdown of rocks in temperate zones will commonly contain equal proportions of smectite in combination and kaolinite with further free illite and chlorite. Areas where rainfall exceeds evaporation tend to produce predominantly aluminium-based clays, particularly halloysite, one of the kaolin group which breaks down irreversibly, losing water and being responsible for surface stabilization of some types of soils. Kaolinites have a general family formula based on $Al_2Si_2O_5(-OH)_4$. However elemental content, as expressed in such chemical formulae, is not a guide to practical usage and material behaviour. The internal arrangements of radicals in the molecule and the extent of water of crystallization in the structure cause major variations in the characteristics of materials. In practice the conservator will be guided more by geological nomenclature and character than by chemistry expressed as collective formulae.

Because of the nature of weathering and wide variations of climate in geological time,

minerals have geographic distributions in which particular clay types are predominant. In humid and tropical areas in non-saturated ground kaolinite is generally found, whereas in cooler and wet climates in well-drained land fine micas, such as illite and halloysite/smectite groupings, are the norm. While broad generalizations of this sort are possible the weathering conditions and the rocks from which the clays take their origins have an over-riding effect on the soils formed. True clays are preponderantly hydrated aluminium or magnesium silicates varying in their ratios of aluminium and silicon. Of the silicate clays, (which are the largest group) one stage of weathering is represented by kaolinite. This decomposition takes place under acidic conditions as a result of the removal of the more soluble metal ions, potassium, sodium, magnesium and calcium, which can be drawn out of feldspars and other basic minerals. The recrystallization of aluminium and silicon compounds forms the kaolinite which in turn can decompose leaving aluminium oxides, iron oxides and silica, which are the basis of the tropical lateritic soils.

The formation of clays is not necessarily a progressive change from an ordered kaolinite to the less ordered halloysites and montmorillonites, although this process is the norm. Laterites may derive from feldspars in volcanic rocks without passing through the intermediary stage of kaolin.

At the upper end of the chain chlorite, which is rich in magnesium, and illite are derived from initial weathering while the smectite groupings of montmorillonite occur at the intermediate stage. The initial sources of clays are feldspars, micas, pyroxenes and others which by relatively moderate chemical alteration decompose to allow their basic constituents to recrystallize as silicate clays. Muscovite, for example, is a rigid non-expansive crystal which is a first product of decomposition from igneous rock. It occurs in granite, gneiss and schists and is the mica of commercial utility. However, it is found in many forms other than sheet. In decomposition it loses some potassium and gains water within the lattice to allow it to recrystallize as an illite, which is a semi-rigid crystalline structure of alternating octahedral and tetrahedral plates which stack into clay particles.

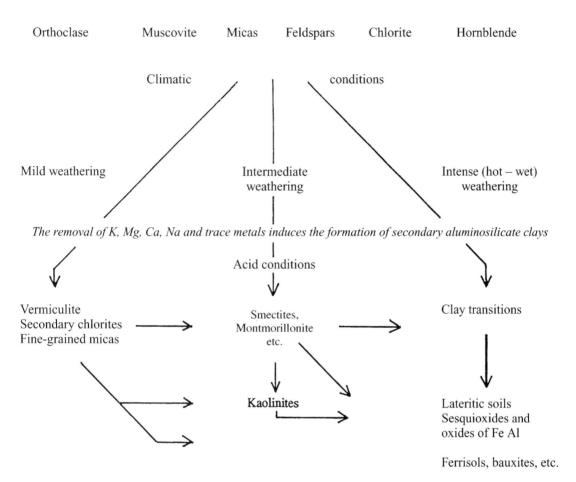

Principal silicate minerals deriving from the weathering of igneous rocks

Orthoclase	Muscovite	Micas	Feldspars	Chlorite	Hornblende

Climatic   conditions

Mild weathering   Intermediate   Intense (hot – wet)
                               weathering   weathering

The removal of K, Mg, Ca, Na and trace metals induces the formation of secondary aluminosilicate clays

Acid conditions

Vermiculite   Smectites,   →   Clay transitions
Secondary chlorites   →   Montmorillonite
Fine-grained micas   etc.

                               Kaolinites   Lateritic soils
                                                            Sesquioxides and
                                                            oxides of Fe Al

                                                            Ferrisols, bauxites, etc.

Figure 2.2 General pattern of weathering, producing clays and associated minerals.

The internal superficial area of smectites is very much greater than those of the other clays and they also have a high capacity to change cations – potassium, magnesium, calcium, aluminium and hydrogen. By comparison with kaolinite they are some 15 times more reactive. This smectite group, which contains montmorillonite, nontronite, soponite and others, swells by the attraction of water into the space between the layers which forces them apart. The plate structure of montmorillonite is held together by weak oxygen links and linkages between exchangeable cations.

A comparative summary of clays is given in Figure 2.3.

The pore structure of dried earths depends greatly upon the nature of their constituent materials, but also relates to the conditions in which they are laid down. The availability of water to act both as a lubricant and to provide forces of cohesion due to surface tension, the presence of mechanical agitation or vibration and simply physical compression all affect the way in which the particles arrange themselves. While most particles are irregular and lacking pronounced directional qualities, organic materials tend to be fibrous and therefore long and sinuous and very much larger than the other particles; and, in brick earths, the clay components tend to interleave or stack by

General Grouping and Characteristics of Clays and Associated Microscopic Minerals

Clays are chemically active through ion replacement: i.e. one element may supplant another in the crystal lattice (isomorphous substitution)

Kaolinite clays

Kaolins consist of stacked alternate 'silica' and 'gibbsite' layers bonded by weak electrostatic (van der Waals) forces. The stacks can be of many thousands of layers.

The type of kaolin derives from variable positioning of the layers due to the several possible forms of interlocking, and from inclusion of varied cations.

Unsatisfied valencies on the surfaces result in **adsorption**, the phenomenon of surface bonding with available ions or with water (i.e. hydroxyl ions which can be driven off at 105°C).

Kaolins are minimally expansive, with the exception of halloysite.

Principle types are:

Serpentine
Kaolinite
Nacrite
Dicktite

which are plate crystals, and **Halloysite**, which is amorphous usually found in hydrated form, often as tubular particles or rods.

Typically, kaolinites are two-layer minerals having alternate silica sheets.

Breadth of crystals typically 1.5 μm.

Specific surface up to 15 m²/g.

Therefore:
minimally expansive.

Typical empirical formula:

$Al_2 Si_2 O_5 (-OH)_4$

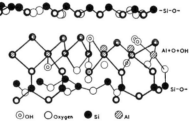

Key-: ○ o   ◉ OH   ● Si   ⊘ Al

viewed on to the plate (above) and from the side (below)

⊚OH   ○Oxygen   ● Si   ⊘ Al

Structure of Kaolins

Smectite-type clays
(montmorillonites predominate)

Smectites are typically three-layer minerals with silical sheets on both sides of gibbsite or brucite sheets. These triple layers form stacks in the same way as in kaolinites.

Due to weaker adhesions between the layers, and strong internal attraction to the hydroxyl ion, montmorillonites are readily invaded by water and are therefore extremely expansive.

Sodium montmorillonite may carry over thirty times as much moisture a a void ratio of 25% in surface pressure conditions as it does under the confining pressures of heavy overburden where it has a void ratio less than 10%.

Specific surface up to 800 m²/g.

Principal types are:
Talc
Pyrophyllite
Muscovite
from which others, i.e.
Montmorillonite: sodium-rich
Saponite: magnesium-rich
Nontronite: iron-rich
Hectorite: lithium and magnesium rich
Beidelite: aluminium-rich
are derived.

Powerfully expansive

Breadth of crystals typically 0.75 μm Several hundred molecular layers of water can attach to clay plate surfaces.

Vermiculite(s)

Dominated by magnesium ions Layers more rigidly structured than montmorillonites.

only **moderately expansive**

Crystal breadth typically 0.3 μm.

Figure 2.3   General grouping and characterstics of clays.

Typical empirical formula:

$Al_2 Si_4 O_{10} (-OH)_2$

Structure of montmorillonites
viewed from the side

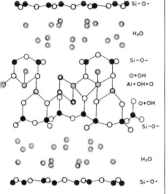

Diagrammatic side-view of the silical layers, top and bottom and gibbsite centre, with water interposed.

Non-clays

Mica(s)
Micas are not strictly classifiable as clays but very small micas are structurally similar, invariably intermingled and in practice are treated as clays. Fine-grained micas are typically muscovite and illite.

Illites
Found in intimate association with smectite clays. Typically very small and alkaline. They are less expansive than montmorillonites due to the greater stability of the layers.
Fine-grained micas – illites – are a derivative of mica by leaching out of potassium.
Crystal breadth typically 0.5 μm.
Specific surface 80 m²/g.

Chlorite(s)
Structurally related to micas, the chlorites are crystals of alternate layers of calcite and brucite. They are equally fine-grained and derive from micas. Despite the name, the element chlorine is not a component. Formed when aluminium hydroxy polymers move into the layers of smectite and vermiculite crystals, they have a lesser capacity for ion exchange.
Crystal breadth typically 0.5 μm.
Specific surface 80 m²/g

Carbonates
Chiefly calcium carbonate derived from precipitation and fine-ground life-shells (e.g. foramaniferae and coccolithophores). It may occur in microscopic crystalline form in intimate admixture with clays, providing material for their decomposition.

Allphane(s)
Amorphous silicate minerals found widely in volcanic soilds.

Laterites are ranges of clays from which heavy weathering has dissolved much of the silica layers. They therefore vary widely on the basis of the parent clays and typically are rich in hydrous oxides of iron and aluminium.

Typical empirical formulae:

$KAl_3Si_3O_{10}(OH)_2$

$Mg_5 Al_2 Si_3 O_{10} (OH)_8 + Fe$

(varied incorporation of Mg, Si and Al with Fe makes up the range of ferro-magnesium silicates)

$Al_2 O_3 2(SiO_2 H_2O)$

$CaCO_3$

virtue of their shape and in consequence of their weak mutual electrostatic attractions. A useful, if simplistic, analogy is of domestic plates or platters in water. Their 'shape' disposes them to parallel or quasi-regular arrangement. Chemical valence or polarity superimposed on this natural tendency causes the plates of clay particles actively to seek a coherent formation. To move into the most logical assembly, however, an input of energy is required. This can come from vibration, pressure of molecular movement or free movement in a liquid where friction and the effective weight of the solids is reduced. In addition, a liquid such as water facilitates the easy rearrangement by acting as a lubricant. Thus, an assembly of plates in clays which might not be achieved in semi-dry conditions by the input of vibration energy will be achieved efficiently if the plates are floated in water. The purification of clays is universally carried out in suspension in water and their compaction generally so. The particle size of clays is very small and the plates can remain in suspension for considerable periods, particularly if the water is physically energized by movement. The particles are sufficiently small to be susceptible to the effects of the molecular agitation of the water. They are not, however, so small as to be sustained in place by this molecular bombardment. If this were so, they would remain in suspension as minute solids and so form an emulsion. That this is not the case is demonstrated by experiment. A mixture of clays and silts may be suspended in water by agitation until it is uniformly distributed and the material is of the consistency of a thin cream. The solution is then continuously introduced at surface level into a stream of clean, deionized water passing down a deep channel formed in a transparent material, the channel being sufficiently deep to ensure that the flow of water is even and very slow. During its progress down the channel the particles fall out of suspension at rates dependent upon their size and shape. In effect, each particle performs a trajectory of fall directly related to its own characteristics and, therefore, the deposited material at the end of the experiment is distributed along the bottom of the channel on the basis of size and nature of the particles. An apparatus can be constructed so that the relative volumes of

particle size can be analysed visually, microscopically or by weight and their percentages estimated. The sequential sizes of particles will be determinable by extrapolation between the extremes. At the upper level there will be larger particles (fine sands), and at the lower level clays. Some organic materials will float out of the analysis together with dissolved and some colloidal material.

Experiment illustrates one way in which clays are laid down in nature, and is exactly the way in which kaolins may be separated from quartz and mica in the production of China clays.

A much less sophisticated, but more practical version of this analysis, usable as a field test, is simply to shake a volume of the earth in a tube with sufficient water (four to six times the volume of earth) to allow total disintegration, and then to set the tube aside for a few hours. A ready guide to the proportions of particles is gained by analysing the layers in which the sediment accumulates. The test is, however, somewhat crude because each particle starts its trajectory at a random height in the mixture, although the rates of fall are sufficiently different to allow the faster-falling materials to precipitate while the lighter particles remain in suspension.

A slightly more sophisticated version of this test involves shaking the earth mixture vigorously in about four times its volume of water with a small quantity of detergent. Immediately the liquid is decanted, any residue being the aggregate content; after a further 30 seconds the liquid is again decanted, the residue being the sand content; after 30 minutes the liquid is again decanted; the residue is the silt content. After 30 hours the liquid is again decanted and the residue is the bulk of the clay content. The entire clay content may be retrieved by evaporation or by longer standing (see also pp. 106–7).

One complication in these apparently simple experiments is the phenomenon of flocculation. This is a condition in which solid particles in a liquid produce a pattern of attraction by their polarity in the liquid, causing them to form fluffy or gelatinous masses, taking up enlarged spaces. The effect is produced by changes in the electrostatic properties of the surfaces which cause them to move apart while remaining in linkage.

Flocculant material remains in suspension as a cloudy mass, rendering the deposition tests ineffective. Flocculation, of which there are two forms, is likely to occur at a median pH value in the order of 5.5 common in soils in suspension. It can be countered by changing pH values or changing the temperature.

2.3 The composition of clays

Earths are generally composed of particles of compounds which are fully oxidized, but occasionally in anaerobic conditions of sustained organic acidity and wetness they may be reduced (deoxygenated) and in consequence a reversion occurs when the soil is subsequently exposed to the air.

While the sandy components of earths are lumpy and irregular in shape, visible to the naked eye and large enough (in the order of 1 mm) to be readily felt, the clay particles are too small to be visible individually or felt. They are regular in form and arrangement and are manifest under very high magnification, typically as tiny semi-transparent flakes, 0.2 μm broad, hexagonal in form and assembled in small stacks or columns, typically some several hundred flakes thick. These blocks form by the mutual attraction of their flat surfaces. Potassium and sodium ions readily associate with these surfaces by the attraction of weak electrical forces, and where this occurs they hold the surfaces slightly apart. Other molecules likewise lodge in these spaces – iron is a particularly common intruder, imparting brown and greenish colours to the clay. Thus, water has a ready access to the spaces in and between the flakes, or wafers, where, in sufficient volume, it can neutralize the weak attraction of the surfaces by forcing them apart. The effect is due to the very short distance over which the forces of attraction are effective. This occurs in situations where the material is fully saturated. However, some forms of clay have particularly strong mutual attractions, sufficient to maintain cohesion even when fully saturated.

Combinations of illite and montmorillonite form strong, stable relationships. These similar, self-attracting, self-assembling flakes are the principal constituents of most clays and readily assemble in alternate layers. In this condition they have a reduced polarity. The form in which this assembly takes place is that of a sequence in which one flake of illite is sandwiched between two of montmorillonite. Collectively, montmorillonite/illite combinations are universally calcium-rich and as a consequence they tend to remain coherent while absorbing large volumes of water between the plates and are the most powerfully swelling clays. Vermiculite, chlorite and kaolinite are much less powerful in this sense. Most clays contain several clay components in varying proportion with correspondingly varied characteristics.

Since most natural clays are mixtures of these primary clays they expand and contract in response to the availability of water. The surfaces of the flakes of clays such as smectites have a negative polarity in the presence of metallic ions, such as calcium, sodium and potassium. Water in the form of hydroxyl ions (–OH) is consequently readily able to move into the spaces between the flakes, causing expansion on wetting and contraction on drying.

Among the strongly swelling clays this effect is more pronounced in montmorillonite than in illite. Illite expands and contracts rather less because it associates in its interplanary surfaces primarily with calcium ions. The flakes are more strongly attracted, surface to surface, and in consequence the absorption of water between them is inhibited. In a typical brick earth containing a mixture of these materials the reduction in volume approximates to a change in linear surface measure of between one-twelfth and one-fourteenth. In a pottery clay it may be considerably greater.

In China clay (kaolin) there is a strong cohesion between the flakes. In consequence, water has less ready access between them; there is minimal free polarity to attract the hydroxyl ion and therefore there is much less swelling and less plasticity. This clay remains relatively stable in fully saturated conditions and requires mechanical energy to shake the components apart. When subjected to such treatment the flakes disperse. Subsequently the material can flocculate (a condition of coherent disorder) and is capable of orderly reassembly.

When wet, all clays are slippery because initial water acts as a lubricant, allowing the

Earth constituents: found in intimate mixture and varying proportions. The differentiation in typical size is a consequence of the processes of the breakdown. Quartz is sufficiently hard to survive typically as grains, whereas other material breaks into smaller particles, and clays, being reformed chemically, are collections of minute crystals.

	Sizes generally	**Typical shapes**
Sands: largely silica (SiO_2), a very hard material, broken down primarily by physical action: surfaces often conchoidal: tend to survive as larger particles.	0.06–2 mm ($\times$ 10)	
Silts: mainly fragmented rock particles, generally less coherent than silica (feldspar mica, etc.), some crystalline material: tend to form fine particles.	0.002–0.06 mm ($\times$ 10)	
Clays: very small components formed and reformed by **chemical action** consisting of stacked plate crystals held in position by electrostatic surface forces: restricted by their nature to microscopic size.	0.002 mm ($\times$ 10 000)	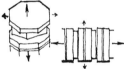
Crystals: principally calcium carbonate (chalk/limestone), calcium sulphate (gypsum) crystallized out of solution: can be silt-sized, or larger, but normally microscopic.	0.002 mm upwards ($\times$ 100)	

Organic materials: (a) colloidal or dissolved residues of life forms, such albumens, organic acids, natural polymers in amorphous condition.
(b) parts of life-form alive or dead – rootlets, bacteria, fungi – formed of lignin, cellulose, proteins, amino acids, etc.; vary in type with climatic circumstances: (a) glue-like or (b) fibrous.

0.002–4+ mm ($\times$ 100)

Water: (a) as water of crystallization found locked up in hard structures such as gypsum: non-available.
(b) bound intimately into ultra-small spaces; e.g. intramolecular zones within the plate structures of clay: non-available.
(c) pore water, as condensate below Rh1 and as vapour, retained by van der Waals forces within even the smallest micropores and attracted most strongly to the ultra-small spaces: partially available.
(d) as a liquid body primarily located at the points of contact between particles, where it has a binding effect due to capillarity but ultimately filling all spaces: available, containing ions in solution

the hydroxyl ion (–OH) where the bond alternates extremely rapidly leaving the molecule H_2O momentarily as the ion –OH.

Figure 2.4 Soil components – typical forms.

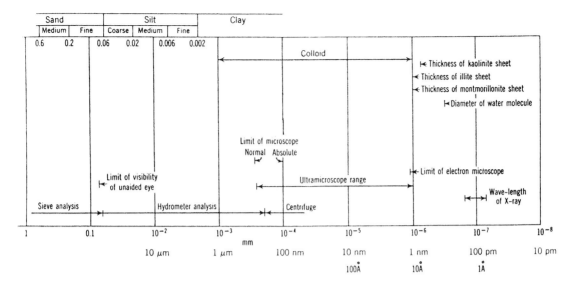

Soil particles: relationship between categories of measurement of size, techniques of analysis and visual accessibility.

Sieve analysis involves passage of particles through a square mesh. The dimensions are given by centres of the divisors, i.e. on-the-square. The longest dimension of a particle may therefore be the diagonal (1.414 × mesh size) as a cross-dimension of an even longer particle which has 'dived' through the hole. Hydrometric sizing is achieved by measuring rate of descent through water and equating the particle to the diameter of a smooth sphere of equal density.

Figure 2.4 Continued

flakes to move easily across their mutual surfaces. When dry, however, the areas of mechanical friction are large and, consequently, a firm though brittle structure is produced. The stability of dried clays is partly attributable to the formation of microscopic crystals, particularly of calcite. This combination is weak in tension despite the coherence between the particles themselves. Weaknesses arise from the large number of minute internal fractures that occur during shrinkage; the material literally pulls itself apart to form minor fissures throughout the structure, and these minor fissures are themselves a part-characteristic of the porosity.

The sheets forming the clays crystallize on variations of tetrahedral or octahedral systems in combination, being either predominantly silica-based or predominantly alumina/magnesia-based. When combined these platelets overlay each other: a typical silica clay has planes of tetrahedral sheets sandwiching an octahedral sheet, each sandwich being interleaved by a less rigid layer in which oxygen and hydroxyl ions carry absorbed cations. These layers form slip planes when the material comes under pressure. On drying, water particles move out from the spaces between the particles, first causing the interparticular spaces to take up water from between these layers. Both these processes involve shrinkage, as the particles and the planes are drawn together by surface tensions and reductions in internal pressure.

Other than the clay silicates the principal groupings at the lowest end of the scale are allophane and similar amorphous alumina silicates, together with iron and aluminium oxides and hydroxides – gibbsite, geothite and colloidal particles with organic components derived from humus. All these particles, even when comprising several hundred platelets, are well below the size at which they are visible to the eye or individually under normal optical microscopy.

The colloidal properties of clays derive from the smaller particles, generally below 1 μm in diameter. The clay fraction between 1 and 2 μm does have some colloidal properties, although the principal amorphous formulations are derived from the smaller particles. The larger particles tend to be self-neutralizing electrostatically. In a colloidal arrangement the dispersions are greater and particle arrangement is more flexible than in crystalline structures. When clay particles are in close association they affect each other by forces which tend both to repel and to attract. The cations in the double layer of clay particles produce a positive electrostatic field which is balanced by a negative charge on its opposite side, expressed on the external face of the particle. Two particles which are then brought together repel each other as a result of the interaction of their superficial negative charges. As the intensity of charge in the inter-layer can vary with the movement of cations, so the external surface charge varies and the repulsion towards adjacent particles varies also. However, other forces also act between the particles – primarily the van der Waals forces which act at the surfaces of adjacent particles. Those forces are independent of the intervening water. Finally there are forces produced by local conditions – hydraulic, hydrostatic and pressures of loading. The net combination of forces determines whether the particles will attempt to move apart or move together, in which case they can behave colloidally and can flocculate. Under loading the forces of attraction and repulsion are unimportant and contact stresses between particles of soil can be very large. Point loads can be of the order of 5–$10\,000$ kN/m^2 over the very small contact areas. The values recognized in construction represent the localized contact forces averaged over the entire area, the large part of which is contact-free. The more densely compressed the material, the greater the frequency of contact points and the higher the overall compression stress.

Much of this interaction takes place between the large particles, silts and sands, and although they are not as complex in physical structure as clays they are by no means superficially simple at molecular scale. However, they are structurally coherent and do not have that capacity to deform in planar directions which is so characteristic of clays. An understanding of their behaviour therefore can be more directly related to the behaviour of materials at sizes which are normally comprehensible. By analogy, the relationship between particles in soils is like the relationship between objects the size of ships and aircraft, intimately mixed in immense number with objects the size of vehicles, mixed in even greater number with objects the size of hay bales, footballs, tennis balls and golf balls, mixed in even greater number with objects the size of books, car tyres, packets of biscuits, compact discs, broken glass and microchips, in volumes whose density might vary as between the heaps in a scrap yard and compressed scrap baled for transport. Every surface is in a continual buzz of atomic vibration as though it were a running engine: every surface is sticky with the glue of electrostatic attractions or slippery with the grease of electrostatic repulsion. Although the analogy is fanciful and inaccurate, it is by such comparisons that a visual understanding of the complex nature of soils may be made graphic. To pursue the analogy further, the growth of crystals might be imagined as the growth of bamboos and fir trees sprouting out in any direction regardless of light or gravity, while the presence of long chain organic molecules such as synthetic polymers would be seen as lengths of rope, string, spaghetti or seaweed carried by flood water into the mix where whales and seals might be taken to represent some of the smaller bio-organisms. This wild analogy seeks to give relative scale to the soil mix. Actual scale may be visualized (with difficulty) by remembering that individual molecules would be visible to the naked eye only on magnification, which would enlarge the head of a pin to between 3000 and 5000 miles diameter!

Most earths used in building throughout history have been essentially composed of quartz, feldspars and clays. Sometimes calcite has been present and can be a substitute for clays. Organic material has often been included but is not essential and is not necessarily a source of additional strength. In earths strengths may, however, be thought of in two senses – the strength in a dry rigid state which is the sense in which it is normally used, and in which it is used here. The term *strength* is

Quartz is the commonest and most durable of all small soil components.

Single silica tetrahedron

Tetrahedra arranged as quartz

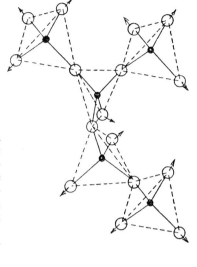

The silicon–oxygen relationship shown diagrammatically. A linkage occurs in which four oxygen atoms embrace one silicon atom. This forms the silicon tetrahedron. This tetrahedron has unsatisfied valencies in this isolated condition. The valency requirements of a single tetrahedron (SiO_4) are satisfied when tetrahedra are joined at the oxygen linkages and a stable structure is produced extending indefinitely, forming a spiral chain.

In its common form the mineral is quartz. In other arrangements it is known as tridymite and as cristobalite.

The regularity of the quartz structure provides its crystalline qualities. The interlocking and slightly twisting sequence of the structure accounts for its rigidity or hardness and hence its predominant survival as larger particles in soils.

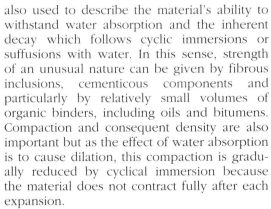

Figure 2.5 The silicial tetrahedron.

also used to describe the material's ability to withstand water absorption and the inherent decay which follows cyclic immersions or suffusions with water. In this sense, strength of an unusual nature can be given by fibrous inclusions, cementicous components and particularly by relatively small volumes of organic binders, including oils and bitumens. Compaction and consequent density are also important but as the effect of water absorption is to cause dilation, this compaction is gradually reduced by cyclical immersion because the material does not contract fully after each expansion.

The densities of soils depend on compaction and on the nature of the particles. Generally, the greater the range of particle sizes, the greater is the maximum density; at the other end of the scale, the smaller the range of particle sizes and the smaller and more angular the particles, the less the minimum density. Such materials will create a stable but looser arrangement of particles. The density of materials can also be described in

absolute terms as specific gravity, i.e. value related to a unitary value for water at 4°C. The general value for soil minerals ranges between 2.3 and 2.9 specific gravity.

2.4 Shrinkage: plastic and liquid limits

The characteristics of size distribution of particles in a soil can be a revealing index both of its characteristics and its origins. If a particle analysis by size is plotted graphically, a curve will be produced characteristic of that material and since there is a very wide range of possible curves, any particular soil will have a distinctive profile. This feature may become a sophisticated analytical tool in determining sources of material. As soils engineering advanced in this century, it became evident that common standards and common methods of analysis were increasingly important in terms of soil behaviour and in 1948 Messrs Atterberg and Casagrande set out standards of

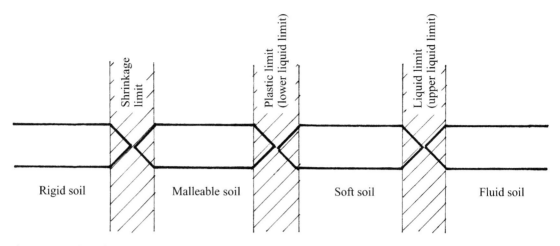

Phases in soil performance with varying availability of free water. The subdivisions of the behaviour of a fine soil with increasing moisture content are defined as the Atterberg limits. At the **shrinkage limit** the moisture content becomes sufficient to penetrate the pores. The soil becomes malleable; friction between particles remains the mechanism of stability. At the **plastic limit** the soil becomes soft or plastic, a 3 mm thread is bendable, dependent on its cohesiveness. Stability derives from the meniscus at points of contact. At the **liquid limit** the water content eliminates the cohesion and particles can move freely. The liquid limit is determined by use of a Casagrande device which analyses fluidity on the basis of tremors required to close a groove of specific width and depth.

The Atterberg limits are a reliable index of soil performance of particular use in soils engineering and of value in work on earth structures.

Figure 2.6   Atterberg limits.

measurements which have become the accepted indices of soil behaviour. They are known as the Atterberg limits and are based on the notion that a soil can pass through four states depending on the percentage of water within it. Between these states are broad boundaries which can be defined and measured by water content. On this basis a soil is said to be solid above a boundary known as the **shrinkage limit**. Above the shrinkage limit it has become rigid, although it will contain water. However, it will not shrink, as opposed to dilate, with water loss, beyond the shrinkage limit. Oven-dry soils placed in normal air conditions will absorb moisture from the air to between 2 and 5% of their content but even in some cases to as much as 20%.

The essence of the shrinkage limit is that it is the boundary below which the effects of water pressure can be observed in terms of expansion. (In this description the term *below* correlates with the *increase* of water content.)

Below the shrinkage limit a soil enters the semi-solid condition. The shrinkage limit is the point at which water has been absorbed into the pores of the soil sufficiently for water pressure to overcome interparticle friction and to allow it some malleability. In this condition rolls of soil 3 mm in diameter can readily be formed and will remain stable.

With increasing water content the soil passes the plastic limit and enters the plastic state. In this condition it is unable to sustain loadings. The **plastic limit** may be defined as the point at which rolls of soil, 3 mm in diameter, fail to cohere. With further increase in water content the soil softens progressively until it reaches the **liquid limit**. Beyond the liquid limit the soil will have entered a liquid state in which the particles have no significant cohesion. The liquid limit can be measured in an apparatus known as the Casagrande device. It is determined by the ability of divided segments of a soil sample to move together under vibration; the number

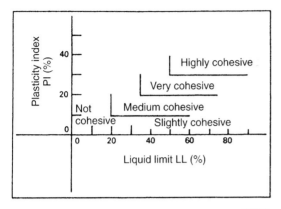

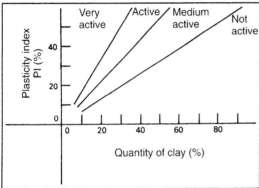

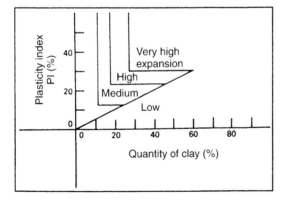

Plasticity indices derive from the Attenberg tests and are calculated as the value of the **plastic limit** minus the **liquid limit** (P1 = P1–LL)

The plasticity index of soil defines usefully its cohesiveness, expansion and 'activity'. These factors are of direct value in determining soil behaviour. The mineralogy of the clays and their proportionate quantities are the joint determinant of soil behaviour.

Cohesion, which can be very high in soils with small particle size, is a factor related to size and separation of cracks on shrinkage.

Activity is defined by the percentage of clays less than 2 µm diameter and is not necessarily directly related to performance.

Expansion (and therefore contraction in 'active' clays) relates to clay type; the individual analysis is specific to a soil.

Figure 2.7 Behavioural parameters of clay soils.

of vibrations required to close the division are counted under specifically controlled conditions. It will readily be appreciated that such determinations are not precise but test results averaged over a significant number of examples provide a consistent pattern of the characteristics of the particular soils and are key parameters in soil engineering. While they are of little direct value to the conservator working on a small scale, they provide a method of understanding the behaviour of large-scale structures and guide the formulation of remedial soils on a comprehensive analytical basis.

The shape of the particles is crucial to the behaviour of earths and it varies across the range. Sand particles are necessarily irregular. They are normally of broadly similar dimension whatever the axis chosen – that is to say, on average the shortest and longest diameters across a sand particle will be defined by a multiplication factor less than five and in the majority of examples, less than three; often less than two. Sands are frequently rounded by attrition in water or by wind carriage.

This is less true of silts where crystalline shapes have been more readily preserved and, therefore, particles with longer axes are more common. Nevertheless, the processes of impact and fracture do not generally allow the survival of particles with very large size differences across the opposing axes. However, below the range of silts the predominating characteristic of clay is a composition of thin, flat particles usually referred to as plate-like, although this is not to be taken as necessarily indicating a roundness in form. Being fundamentally crystalline, their edge profile tends to be angled, typically with hexagonal profiles. Seen under a microscope these particles are more comparable to pieces of slate which cleave readily in one direction.

Clays are not formed by attrition of larger particles, they are formed by physical forces and chemical reaction. They are then transported by water or wind. More broadly speaking, *clays* as a size definition may be a mixture of clay minerals and very fine non-clay minerals. The particles are not visible to the naked eye.

2.5 Water

As water is a fundamental constituent of all assemblages of earth particles used in building work and as the behaviour of water varies in relation to particle size, polarity and shape, almost no consideration of the processes of manufacture, use, weathering and conservation can be divorced from the effects of free and combined water.

Water is attracted into very small spaces. It therefore exists in soils in the free (liquid) form, as captured water in interparticle spaces and as physically engaged water locked up as water of crystallization. The spaces left between particles are a function of their sizes and shapes, the processes of adhesion and compaction and the presence of organic matter which is rarely absent and may provide as much as 5% or more by weight of the content of a soil mixture which,

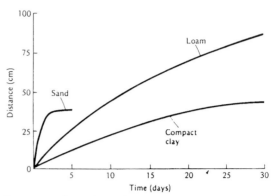

Water rise through earth structures is measurable in both rate and height – the smaller the pores the slower the rise. Height is determined by pore size and nature, and also by the rate at which water is removed.

Figure 2.8 Rise of water in earths.

due to its lesser density, may be in the order of 10% by volume. As organic matter, however, produces large molecules and as these are assembled into many types of fibrous structure, the effect of organic solids is out of proportion to their component value. Organic components are almost always the product of plant or animal life and their effect is universally apparent in earths. Even in fired brick their presence will have affected the arrangement of the particles prior to heating and so have partially determined the shape and nature of the spaces between the particles.

Water is always present in the pores of any form of masonry, having a special affinity for the smallest of intermolecular spaces. Water is a universal component of living organic structures: it is bonded chemically into the molecular relationships and into the tissue. Physically it is also bonded as water of crystallization into the crystalline structures of earth particles and it occurs as a free solid, liquid or vapour within soils. Water can constitute as much as 50% of the volume of saturated samples and even the most densely compressed naturally occurring sample, a mud rock or shale, will contain some 20% by volume of water when fully saturated. Sands, whose predominant silica particles do not

A simple linear formula describes capillary rise in a tube

Height of rise = 2 (surface tension × cosine of the angle of contact α)/d × r
(cosine must be equated for the angle above 90°)

where d = the density of the liquid and r = the radius of the tube. A calculation in soils is similar but more complex due to other effects at small pore sizes (see van der Waals effects and Poseuillès theory). Determination of the angle of contact (α) is not straightforward, but valid assumptions can be made where the angle is between 60° and 120°.

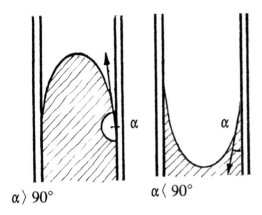

α 〉 90°

Hydrophobic, where surface attraction repels liquid (as with silicone coatings)

α 〈 90°

Hydrophillic, where surface attraction causes liquid to rise, (as with silicaceous surfaces)

In practical terms in soils rise is determined by experiment.

Measured capillary rise in free-drawing soil structures

Soil	Void ratio	Max. capillary head (mm) Saturation capillary heads are about half these values
Sandy gravel	0.45	28.4
Fine gravel	0.29	19.5
Silty gravel	0.45	106.0
Coarse sand	0.27	82.0
Medium sand	0.48–0.66	239.6
Fine sand	0.36	165.5
Silt	0.95–0.93	359.2

Figure 2.9 Capillary effect.

absorb water, will provide space between the particles of some 35% of its own volume.

Liquid water is a powerful solvent. In solution many compounds – or atoms – are ionized, that is to say they carry an electric charge due to loss or gain of an electron or electrons, usually in the process of subdivision. Cations are positively charged and anions negatively charged. Typically ions arise in solution by the subdivision of a salt, e.g. sodium, chloride (Na_2Cl) will provide a cation Na^+ and an anion Cl^-. The importance of recognizing this distinction lies in interpreting chemical activity.

Water within soil falls into six categories:

1. Vapour within the air spaces: this is important because, as vapour, the molecules travel fastest and furthest. This water is contained within the air of even dry soil.
2. Freely moving water: this is the liquid body which flows and is continuous in the larger spaces within the soil. It is linked by a hydrostatic pressure. It may be squeezed out of soil and is affected by capillarity. Wet soil carries freely moving water.
3. Adhering water: this is the body of water which, in the absence of freely moving water, remains on the particle surfaces, coating them and affecting their interparticle behaviour. It is controlled by the surface polarities of the particles. Its effect is apparent in damp soils.
4. Pore water: contained within the soils are very small spaces in which the behaviour of liquid water is governed by capillarity and of water vapour by the effects of van der Waals forces. Such water is regularly present in 'dry' soils and is driven off by drying at close to boiling point.
5. Absorbed water: this is liquid, but contained within the interstices of particles, such as the spaces between the laminae of clay plates. Its retention by polar and ionic forces is such that it is not driven off unless temperatures of up to 200°C are maintained.
6. Water of crystallization: this water has entered into the structure of the crystal lattice, and is no longer liquid. The form of the solid depends on and derives from its presence. It may be driven off (as from gypsum at temperatures upwards of 150°C).

2.6 Composition of earth materials

Magma, which forms the mantle of the earth and from which is derived the material of its crust, is composed essentially of elements of the middle range of atomic weights, producing in turn compounds generally lighter than the core metals on which the mantle floats. These elements have universally combined with oxygen and generally also with each other in relatively complex compounds, often structured as crystals, sometimes by the incorporation of water of crystallization as in clays to provide a regular molecular arrangement which satisfies surface polarities and becomes stable.

Other than oxygen, the commonly occurring elements in this middle range of atomic weights are carbon, silicon, aluminium, calcium, sodium, potassium, barium, magnesium, boron, fluorine and sulphur with crucial proportions of heavier metals, particularly iron. Singly or in combination these elements form the cations which bond with other highly reactive elements (oxygen and hydrogen) to form the complex molecules general in mineralogy. The resultant range of compounds is inherently complex and, by virtue of crystal mineralogy, becomes even more so. This complexity derives at least in part from the conditions of extreme pressure and temperature coupled with long periods of time in which the base materials have been held deep below the earth's surface. Once broken down, many of these complex arrangements can reassemble only by repetition of the same conditions. The process of metamorphosis (cohesion under high temperature and pressure, changing the nature of the rock), which produces, among other things, the marbles, occurs under such conditions. The most telling example of all is perhaps diamond, the hardest crystal, formed from entrained carbon, which takes on a rigid lattice structure but when combined with oxygen by burning disappears into the atmosphere as carbon dioxide. Such extreme differences in form are rare but the immense diversity in minerals can derive from variations which occur in their formation. Comparison of calcium carbonate occurring as chalk, limestone and marble illustrates the point.

Chemically the combination which forms minerals demands the coming together of a cation and an anion, but rather than imagining the relationship in simple terms, as with the opposite poles of a magnet, it should be thought of in terms of a grouping in which the linkages, positive to negative, largely or completely balance others. Thus it is that particular cations can take up several different types of grouping. Two minerals, each having very different characteristics, may be formed of identical ranges of cations and anions. The difference depends entirely on different crystalline or structural arrangements of these components. At its simplest level chemically, the comparison can be made between diamond which owes its intrinsic qualities of hardness and transparent brilliance to its dense crystal lattice, and graphite, where carbon as a plate structure produces a soft black material with totally different characteristics but likewise producing nothing more than carbon dioxide when burned. These varying arrangements of an identical element are described as allotropes. Other variations in the character compounds derive from the compounding of crystal structures, and yet others from amorphous characteristics.

Thus, a relatively simple clay material, allophane, is an aluminium silicate. This material is generally amorphous, that is to say, it takes up no crystalline or rigid structure and is a widespread constituent in soils. It has a high capacity to attract and absorb cations and considerable ability also to absorb anions. In its basic state it tends to be colloidal but with the absorption of available cations it can move into a stage of crystal formation when the same components become locked together into a rigid structure containing very large numbers of molecules. Chemically, therefore, it can remain the same while changing its physical characteristics from being a freely moving molecule to being a component of a rigid structure. As an extension to this concept, the behaviour of water entering into crystal structures introduces further ranges of molecular construction.

Allophane also illustrates an important characteristic of the particles predominant in clays. It is colloidal. In water it becomes a sol, that is to say, it enters into a stable suspended condition in which the molecules join in amorphous extended groupings as very small particles in length occupying a substantially larger volume than the material in a dry state.

Colloidal states are also a well-known characteristic of clays. Due to its plate structure a fine clay has a greater superficial area than the same weight of a medium sand by a factor of upwards of 10 000 and since the behavioural characteristics of these materials relative to water are predominantly governed by surface phenomena, this surface relationship takes on prime importance. The behaviour of clays is partially determined by the relationship of colloidal size particles to the crystalline plates. A relatively simple graphical exposure of the increasing surface area of a given weight of material demonstrates that the increase in surface area is in a ratio with the other characteristics which include cohesiveness, swelling and plasticity.

These qualities are negligible in sands and coarse silts, rising to become important through the general clay sizes and increasing enormously through the colloidal stages. The water-holding capacity of very fine particles increases as particle size decreases and absorbed water in the very fine pore spaces enters into a lower energy state than in the free condition because of the restraint upon molecular movement. This is reflected in increasingly rapid molecular movement, appreciable as an increase in temperature on absorption and a reduction in temperature on discharge of the water from the clay. For this reason an input of energy during the process of discharge facilitates the movement. The physical chemistry of the molecules has another important grading effect. The readiness with which aluminium and magnesium form compounds that enter the fine stage of subdivision causes these elements to become more generally represented in the final particles. Conversely, the resistance to fine subdivision of straightforward quartz sand causes it to predominate in the coarser material. By separating out the finer from the coarser particles of a soil sample, therefore, the chemical constituents are proportionately differentiated. Thus a medium sand may contain 90% silicon oxide (quartz), a fine silt only 65% and a fine clay soil 30%, dependent in part on the conditions of transport and deposition.

By contrast, aluminium oxide may be much less than 5% of a coarse sand rising to between 10% and 15% of a fine silt and 25% of clay. It must not be thought, however, that the chemical representation indicates that the elements occur in separate particles, as can be shown by considering the clays, which are generally alumino silicates, but containing varying quantities of magnesium, potassium, iron and other metals such as calcium and titanium.

A simple representation of the proportionate ratios can be set down in a diagram which shows how the sands are predominantly composed of primary silicate minerals and the clays of secondary silicates. The pattern is completed by recognizing that non-silicate materials tend to fall in the middle of the range, forming a larger component of silts than of sands or clays, but it must be remembered that the pattern has wide variations (see Fig. 2.1).

2.7 Mineralogy and mineral decay

Two systems of mineral definition necessarily run in parallel – the chemical and the geological. In the ultimate analysis the content of a material may be known by the elements it contains, but a description of the elemental breakdown is no guide to the form of the material. The conservator, being concerned with practicalities, will rely primarily on descriptions of a geological type such as limestone for one structure of calcium carbonate and marble for another of virtually identical chemical mix.

Igneous rocks solidifed in the crust from molten magma are composed principally of silicon and oxygen in the form of quartz and feldspars, the commonest minerals in the earth's crust. These occur in various mixtures forming granites – rhyolite, felsite, obsidian, diorite, andesite, basalt and hornblende are among the most common mineral variants. The terms *gneiss* and *schist* are reserved for modifications to these basic rocks metamorphosed under pressure and temperature.

Other important modifications include the deposition of calcite (calcium carbonate, $CaCO_3$) from aqueous solution and marine animal skeletal deposits. This deposit crystallizes to cement particles into rock or to form rock in its own right, giving rise to the ranges of sandstones, limestones and chalks and again, by metamorphosis, marbles.

The inexorable, inevitable, interminable processes of weathering, ultimately destined to reduce this planet to a sea-covered sphere beneath which all solid particles will lie, act primarily by chemical change and by dissolution. The chemical process is aided by the mechanical processes of impact and fragmentation. Temperature, air movement, animal and plant activity, physical movement of ice and water and the immense heat and pressures within the crust cause the materials to disintegrate while remorselessly their constituents change by oxidation, hydration, the actions of acids and basic solvents etching and dissolving until the solid rocks break down. Quartz is the most resistant, followed by muscovite and the feldspars, followed by biotite and hornblende, followed by olavene, dolomite and calcite and softer materials such as gypsum. This hierarchy of resistance to weathering accounts for the predominance of the more resistant materials as soil particles. Each successive product of breakdown offers a new range to assist or promote the next stage of decay. The higher the temperature and the more humid the climate, the more quickly the process goes on until granites become clay and sand and soluble minerals are washed into the seas.

2.8 Clays as minerals

The structure of clays is crystalline and therefore regular, the general form being that of small sheets in combination so that each particle will consist of a multi-layered plate – in simple terms a multiple sandwich – which has a much larger internal surface than its superficial external area. Water molecules comprising anions and cations (–H and –OH interacting) can be absorbed within the internal spaces of the sandwich structure. In some of its most stable forms the clay crystal will consist of sheets of different molecular formation depending on the predominance of silicon, aluminium or magnesium. The sheets do not necessarily adopt identical profiles and a tetrahedral sheet may overlay an octahedral

sheet, forming a complete rigid structure. The specific construction of each of the clays affects their properties to a relatively minor degree but their general qualities – cohesiveness, plasticity, moisture retention, swelling, shrinkage and slipperiness – combine with their capacity to produce rigid, hard materials to give the properties for which they have been so widely used and it is with these properties that the conservator will be primarily concerned. Nevertheless, in practical terms the subdivisions of type can be important in acquiring a balanced mixture and understanding the probable performance of the clay.

Clays divide into the structured and colloidal. In analytical terms, virtually every clay is a mixture of both. The structured clays tend to produce a cohesive mass in which each of the plate-like structures, consisting of multiples of multi-layered molecules in the fashion of a wafer biscuit, will, under the right conditions, orientate themselves coherently. Random in total suspension, they move under conditions of pressure and vibration to take up compact arrangements, whereas the colloidal clays resist coherency due to linkages which hold them in a more distant relationship one with another.

The common clay components are kaolinite, montmorillonite plus illite (also known as smectite) and allophane. These represent various stages of weathering and the types move from one form to another with the repositioning of metallic ions as the processes of leaching and reaction go on. Leaching is essentially a process of washing out, resulting in the transfer of ions from one structure to another, producing a modification of the clay chemistry. Consequently, as these processes continue, one form of clay can be transmuted into another. Kaolinite represents the stage of weathering lower down the scale than illite), vermiculite and chlorite. Illite and montmorillonite appear in such close integration that geologists find it convenient to describe them jointly. Their physical properties, however, are very different: the internal surface area of montmorillonite is some 50 times greater than illite, which itself measures $15\,m^2/g$. The proportion of different components within a mix has a powerful bearing on the behaviour of the earth. Chemical reactions generally

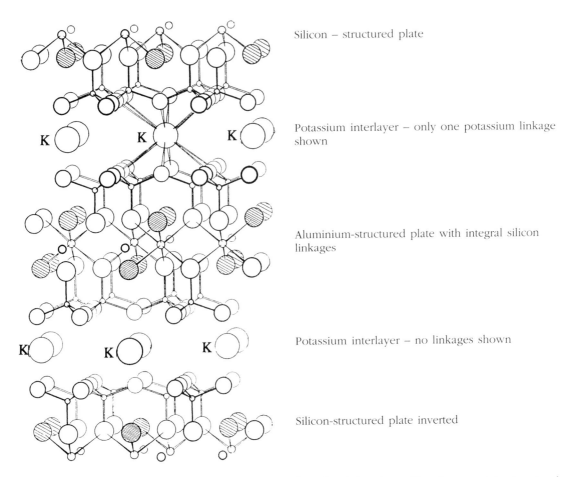

Silicon – structured plate

Potassium interlayer – only one potassium linkage shown

Aluminium-structured plate with integral silicon linkages

Potassium interlayer – no linkages shown

Silicon-structured plate inverted

The lattice structure of a section of a clay seen from the side of the plates. The plates must be imagined to extend towards and away from the viewer (at an angle). The size of the circles is diagrammatic, differentiating the ions – aluminium, potassium, silicon, hydroxyl (shaded) and oxygen. This mineral is **muscovite**.

Figure 2.10 Molecular structures of a clay.

proceed faster in warm and humid conditions and the weathering of clays in the tropics, aided by organic processes, has produced clays which tend to be lower down the scale of degradation than those in temperate zones, with the consequent preponderance of the iron-rich haemetite types of soil and higher proportions of colloids. The result is a soil which is more sticky, characteristically red, and often described as laterite.

The principal groups of clays and associated materials with their characteristics are set out below:

- Smectite group: montmorillonite and others, saponite, nontronite, beidellite, etc.: high plasticity, high shrinkage/expansion characteristics (a smectite lattice). Primary and widely distributed group of clays with very small particle size producing a very hard dried material. Flake size 0.01–1 μm. Internal surface area 500–800 m²/g
- Vermiculites: less plasticity and expansion than smectites; water acts to retain the clay plates. Fine-grained, moderate expansion, magnesium/aluminium-based

- Illites: fine-grained micas, such as biotite and muscovite. Very modestly expansive, flake size 1–100 × smectite, potassium-bound
- Kaolinites: includes kaolin, halloysite, dicktite, nacrite. Non-expansive, modestly cohesive, larger particle size and thickness, minimal interleaving, stable product of extended weathering. Crystals 2–5 μm. Internal surface 15 m^2/g
- Chlorites: ferro-magnesium silicates which generally have similar physical characteristics to Illites. Often a product of leaching of magnesium from weathering of smectites

- Allophane: ($Al_2O_32SiO_2$ up to $5H_2O$). Amorphous, modestly expansive, cohesive, colloidal, widely mixed in soils
- Gibbsite: ($Al_2O_33H_2O$) and geothite ($Fe_2O_3H_2O$): some crystalline structure, not as cohesive as silicate clays, product of extensive weathering, widely distributed, lateritic

These characteristics can assist the conservator in assessing the probable performance of a material of which a geological description has been offered. However, inferences based on generalities must always be subordinate to and supplemented by laboratory or practical testing.

3

Earths as building material

3.1 Mechanics of cohesion

Discounting those earths which contain bituminous materials, there are effectively two types of mix which have over the ages proved suitable for building work – those containing calcite (chalk, lime or limestone) and those without. In the first, crystalline formations provide a mechanism for solidification and coherence and the group could be broadened to include gypsum-based materials where these occur. The mechanism involves the inclusion of available water which enters the compound at molecular level as water of crystallization and becomes bound up in the material itself. Where calcite is not present, although some crystallization can occur, friction is the predominant mechanism of coherence and the tensile strengths of these materials are, therefore, generally lower.

It is common practical experience that where friction is the prime method of transfer of stress, variation in particle size produces the most stable results. This is true of material compacted into foundations and road bases and on a much smaller scale it is true of earths. The mechanics of this phenomenon depend upon a triangular distribution of forces repeated many times over. In rather crude terms it can be thought of by considering three near-spherical stones in contact. They touch at three places only, leaving a substantial void between them. This void may then be imagined as filled with a matrix of similar smaller stones, each of which similarly touches one of the prime stones and two others, the intervening spaces being filled with yet smaller particles in the same geometry. Each triangle of contact defines one plane and, except in a regular crystal lattice, no triangles lie in the

same plane. The triangle is an inherently rigid shape and, although the points of contact being restrained only by compression and friction are not themselves fixed or rigid, the effect is that while the material remains in compression the distribution of pressures can remain sufficient to meet forces of disruption. The multiple triangular matrix transmits between the larger particles a pattern of forces which would not be achieved by the points of contact of these larger particles alone. Forces over the small points of contact can be high.

This overly simplistic explanation offers a method of understanding the greater strengths obtainable in a soil whose particle sizes are distributed across the spectrum of size as opposed to the strengths achieved by a mass of particles of identical size. The forces of friction can be increased by the interlocking effects of crystallization and by the cohesion given by liquids present, including oils and water, whose viscosity and surface tension provide a force holding the particles together. Further binding effects may derive from long chain molecules which wrap themselves randomly around particles. Clays can, under certain conditions, form rope-like and quasi crystal-like structures in addition to behaving as stacked plates which exhibit differential friction – sliding easily in one plane but being very difficult to move transversely.

3.2 Pores and structure

It is a general characteristic of earths that the particles cohere by one or a combination of several mechanisms – surface polarities, friction and the interlocking of fibrous crystalline material. The cohesion provided by

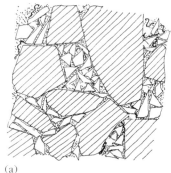

(a)

Diagrammatic cross-section through a small sample of soil in a state of collapse in a saturated and therefore fluid condition. All particles are free to move under pressure. Some air remains entrained in globular spaces between particles. Some organic material remains entrapped.

The triangulation between larger particles has been re-established but without cohesion at contact points due to the lack of any binding material. In addition the particles have lost weight due to immersion, an effect which is accentuated by small bubbles of entrapped air. Despite the triangulation cohesion is therefore minimal.

(b)

Diagrammatic cross-section through a small sample of soil in a plastic state in which the clays have swollen due to absorption of water molecules into spaces between the plates of the crystals. Enough water is available to form large concentrations at points of contact, reducing the effective surface tensions and allowing particles to move apart under the pressures of enlarging stacks of clay crystals. The platelets can also move laterally, contributing to the sliminess of the material. Air remains entrapped in the pores sufficiently to maintain continuous air spaces and this also contributes to the slippery character of the material.

Some water in the soil concentrates at the points of contact where, due to solution, microcrystals will have lost cohesion. There a meniscus forms holding the particles together by surface tension. This fluid bond provides malleability. Other water moves into the interstices of the clay particles (i.e. between the plates) and at their points of contact. The clay particles swell differentially, depending on type. The amount of water available initially, however, is not sufficient to provide for more expansion than is readily taken up in the air spaces of the pores, without overcoming the cohesion given by surface tension, Increasingly this expansion is apparent externally, the physical contacts are less and the earth becomes more plastic.

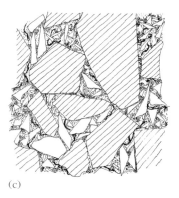

(c)

This diagram shows many three-point contacts. In reality, very few three-point contacts would be visible because the triangulations lie in many different planes.

Diagrammatic cross-section through a small fragment of rigid dried soil. Rigidity derives from the multiplicity of triangulated points of contact. Cohesion at these points may be due to water containing large amounts of dissolved material as the result of evaporation, or microcrystalline or cohesive organic material left when all free water has been removed. Pores are filled with air with a substantial moisture content.

An effect of drying out is to concentrate water at the points of contact and increase its concentration of ions until ultimately dissolved material is left as crystals or 'glue' at these points, providing a basis of coherence.

Note: This earth might contain 55% sand, 25% silts, 15% clay and 5% other.

LEGEND: particles shown diagrammatically.

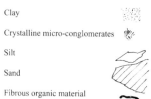

Clay

Crystalline micro-conglomerates

Silt

Sand

Fibrous organic material

Figure 3.1 Soil structure: (a) in a fluid state; (b) physical cohesion between shrinkage and liquid limits; (c) physical cohesion in dry or semi-dry soil.

the surface tension of water is a significant additional though variable force. By contrast, molten components, later solidified into a vitreous continuity, are the essential binder in volcanic stones. This, in part, is effected as crystallization which takes place upon cooling, and partly by vitrification, which in itself has crystalline properties. Igneous rock, therefore, can combine both the 'gluing' effect of solidified liquids of a glassy (vitreous) nature and the interlocking action of the formation of mineral crystals. Being formed at high temperature these crystals are not dependent upon water as a medium for the transport of materials into the crystal or for the integral structure of the crystal itself (i.e. they do not necessarily contain water of crystallization). Earths, being largely mixtures of decomposed magma, contain material of igneous origin either as individual crystals or as conglomerates of these crystals. Finally, some earths are given coherence by oils, fibres or resins, which provide a linking mechanism through the introduction of long chain molecules acting as binders.

In unburned earths mechanical friction, coupled with some crystallization, is the prime mechanism of solidification; the same process of crystalline setting of complex compounds of carbonates, silicates and aluminates is the key to the solidity of mortars. The formation of crystals in close proximity produces an intimate arrangement in which the crystals cannot be separated other than by breaking them.

Continuity between the materials is provided by porosity, which in turn derives from the nature of the particles of the material and their packing. The nature of the pore structure relates to particle size and to the method of adhesion. The crucial factor in determining its behaviour, however, is the nature of the surfaces, for these govern their reaction to water and to ions in solution. These ions in processes of interchange can cause the molecular structure to alter chemically, thereby fundamentally altering the nature of the material.

It is a broad truth that pore size is a function of particle size, but particle shape and arrangement are also significant. The behaviour of the material is governed by, among other things, areas of surface contact between particles and typical distances of separation between surfaces. If the problem is imagined on a larger scale, with particles being the size of pebbles, it is immediately apparent that the larger the stones, the larger the spaces between the particles. If the stones are all rounded their points of contact are minimal and the area of surface contact is small, the pores are continuous and voids are effectively linked. If the stones are more like crushed rock with irregular conchoidal faces, more surfaces lie in close contact and there is a change in the nature of the pore. Much of the void then consists of thin, wedge-shaped volumes in which the surfaces are relatively close and have a much higher proximity: in other words, at any given volume the surface area of the pore is larger. If smaller particles are added, this effect is enhanced. The surface area of the pore increases for any given volume and there is a decrease in the average distance of separation of the solid particles. To pursue the analogy further, particles may themselves be hollow, laminated or perforated, enlarging the surface area many times. Clays, composed of multi-layered rigid plate structures, have enormous internal areas of fixed pore size, accessible to water. On a large scale, the clay particles might be imagined crudely as sheets of corrugated cardboard tied loosely into piles half as deep as they are wide. Their internal surface areas are so great that one authority has estimated the total surface area of the clay particles in a 'furrow depth of a hectare of clay soil' as being 'twice the surface area of the United States of America'. By such analogies it can be understood that a very small pressure exerted within the pores of such material can have a very powerful cumulative effect.

Transferring the analogy to a microscopic scale, a number of conclusions hold good. Pore size and particle size retain a roughly consistent relationship: the surface area increases in a ratio as particle size decreases. The more irregular and faceted the particles, the greater the area of surface contact, the less the average distance of surface separation and the greater the superficial surface of the pore.

In 'settled' arrangements the average pore size in relation to particle size decreases and there is a corresponding increase in the average

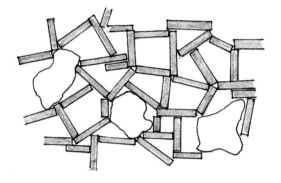

Flocculated deposition in which electrostatic edge effects cause platelets to adhere by edge contact

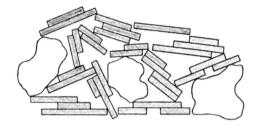

Normal deposition in which electrostatic charges are balanced between edge and face effects

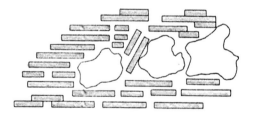

Consolidated deposition in which a force has induced greater packing and hence alignment

Note: clay particles will stack more than is shown in these diagrams.

Figure 3.2   Fine-particle soil deposition.

ratio of surface area to particle size. It will be evident that these conclusions depend upon a coherent arrangement of the particles. In a random and very loose arrangement the ratio between pore size and particle size is exaggeratedly large, while the converse is the case when a compressed and mechanically rational arrangement between particles has been achieved. Colloidal arrangements are such that the particles are held apart and the intervening spaces are therefore proportionately greater.

Pore size has other important ramifications. The behaviour of liquids, particularly water, in small pores varies at a rate disproportionate to the reduction in pore size. In very small pores condensation takes place more readily than in open-air conditions and the ability of liquids to move through the pores is inhibited. This phenomenon is expressed in Poiseuille's equation, where the resistance is defined by an inverse factor of the pore radius to the power of four. In other words, movement is enormously inhibited by reduction in pore size, provided the material remains coherent, despite saturation. It is to this resistance to the movement of liquids that the water resistance of a saturated clay may be attributed.

3.3 Mortars in earth construction

Mortars differ significantly in their chemistry and behaviour. All, however, are porous in general terms and the porosity is similar to that of dried clays. Some cements, however, can be so water-repellent that they become impermeable to liquid water. Limes, cements, gypsums and earths cannot be divided sharply from each other as can baked clays from unbaked earths. They are variably intermiscible and can include, as active (i.e. chemically reacting) constituents, ash, ground ceramics or lava and brick dust. Mortars are working elements of masonry structures and must be understood structurally and chemically as an integral part of a wall mass. Particularly in earth structures they may be stronger then the blocks they contain, e.g. Portland cement mortars. Generally it is an advantage if the mortar is weaker than the block, since it provides a common denominator for the stresses within the wall.

The oldest, most fundamental mortar is earth. Because it is generally laid wetter than the block or brick it supports, it loses more water and is, therefore, more porous, and so tends to be weaker. The character of the earth may be modified by the inclusion of lime, gypsum or organic binders, such as reed or straw, or albuminous products, such as blood or faeces, or modern chemicals such as detergents. It may generally be said of these organic introductions that, although their initial effect may be advantageous, their properties may diminish in the longer term.

Integration between wet earth mortars and dried earth blocks is rarely a problem. Suction and the inherent roughnesses of the surfaces provide sufficient interlock to achieve a full rigidity. In many climates and soils this is enhanced by wetting the blocks prior to laying. The problem with earth mortars is that, being wet and below the liquid limit on placing, shrinkage occurs with setting. In the perpends this shrinkage may be expressed in

cracking within wide joints. This may be important at junctions. In horizontal courses this problem disappears, as shrinkage is taken up by settlement.

Historically many attempts have been made to amend mortars to avoid the problems of shrinkage. Where successful, these have often involved the inclusion of sands, lime and/or gypsum. In modern times cements have been used. The result has been in many cases a mortar stiffer and ultimately harder than the wall material and more resistant to erosion. Where the wall is sheathed and weather-protected the combination has proved durable and efficient, often being compatible with the outer weathering material. Where both earth bricks and a harder mortar are equally exposed the effect is to accelerate the decay of the bricks by wind erosion and by the edge effects, where water is absorbed and discharged readily from the earth at its junction with the mortar, accelerating the breakdown of the earths and rapidly leaving a void as the face of the earth block weathers back.

Limes are the product of calcining calcium carbonate in the form of limestone, marble or chalk and rehydrating the product which is used as calcium hydroxide (lime putty or, dry, as a hemihydrate). Gypsum, roasted at lower temperatures, gives off water of crystallization to become plaster of Paris, which resumes a rigid structure as water is taken up again. Cements set by complex crystallizations, principally of aluminium silicates in wet conditions, and some of those similar crystallizations occur in the setting of hydraulic limes. Apart from some products of organic decay, nothing normally present in earths inhibits any of these setting actions. Mixed with water in the earths, any combination of these setting agents will achieve its set and so provide a rigid basic mortar for the walling. It may be observed that, while limes and cements shrink on setting, gypsum expands. Consequently a judicious blend can achieve a near-stable set. In most mortars which can fairly be described as earths the setting agent may be less than 5% by weight, perhaps bulked up substantially with sand. In some instances much stronger mixes may have been used and compatible additives such as crushed chalk, limestone or gypsum rock may have been included. Some of the hard mortars used in the early Muslim period have included fired and raw gypsum.

Stabilized earths have been used as mortars through long periods of history. Their nature must be understood prior to conservation.

3.4 Swelling and shrinkage

It is in the nature of earths that swelling exerts a powerful force which is complemented on drying by fracturing of the material as shrinkage takes place. The effect can be demonstrated by packing a container with small perforations in its walls with a clay-rich earth in plastic state. The container is then sealed and water introduced through a tube to the core. As water is taken up clay is extruded through the perforations. However, on reversal of the process the cohesive strength of the clay is insufficient to overcome the frictional resistance and the material is not drawn back. Instead it breaks off. This is an extreme example of an important phenomenon which causes damage to porous structures, and particularly to unbaked earths which expand and shrink to a linear dimension typically of one-twelfth or one-thirteenth. The more clay-rich the material, the more cohesive strength it is likely to have. Consequently, on contraction a clay-rich material will be capable of cohering or drawing itself back over a wider area of surface then a more sandy or 'earthy' material. This effect is familiar in the strains seen as shrinkage cracks shown on an area rendered with clay-rich earth. The richer material will retain its integrity across wider zones but the cracks, where they are forced to occur, are then very much larger. The cohesive strength of the material has allowed larger panels to remain integrated during the contraction process, but since the ·overall linear contraction is greater the crack, where it occurs, is correspondingly larger. Conversely, the 'earthy' material will contain very much more frequent cracks but they will be smaller, closer and shorter. This phenomenon is met with similarly in the application of cement and lime renders. The weaker the material, the less visible and preponderant the cracking and the more its flexibility in accommodating strain.

3.5 Freezing and thawing

A further mechanism which can cause internal stress and, therefore, damage to the structure of the porous material is the change in volume which occurs in water at low temperatures. While much of the expansion below 4°C is taken up by the compression of gases or the movement of gas and water vapour through the material, there can be circumstances in which the trapping of water in partially enclosed zones produces an increase in pressure. A freezing of water on the outer perimeter, leaving liquid water at the core, can then produce a condition in which the core, as it freezes, suffers disruptive expansion and damages the structure of the material. The internal stresses may be distributed hydraulically.

3.6 Crystal growth

However, the more significant action of freezing probably occurs in the formation of ice crystals which, by diffusion, draw towards them liquid water which feeds the crystal growth. The growth of the crystal causes an expansive stress and an internal pressure in the material. Similar internal stresses can be created by the growth of crystals other than ice. The pressures of the mechanisms by which ions move into a crystal lattice are substantial. As the lattice forms in an enclosed space, the growing crystal exerts a pressure upon the enclosure which will effectively be a thrust against neighbouring particles. Thus, a porous masonry material containing a solution can, in a drying condition, be in a situation where the ions, once they have been located in the crystal, are no longer free in solution. In consequence, more ions move in and effectively a pressure or stress is created, simply by addition to the crystal structure. Osmotic and capillary pressures aid the movement of molecules into the lattice, and the growth of the rigid crystalline structure impacts upon the adjacent matter, exerting a force against it. In freezing conditions such crystal formation, in which cooled water is supplied to the tip of the crystal, accounts for the long needles of water crystals (ice) sometimes found as pillars in frozen earths, carrying above them the disrupted surface.

Crystal formation can, by this effect, cause porous materials to expand and, particularly, can be the cause of their decrepitation by spalling. The formation of both ice crystals and of the crystals of metallic salts can in this way disrupt the coherence of a porous structure which, in the worst case, can simply collapse as a powder when the crystalline structure decays, thaws or is dissolved away.

Internal damage may have occurred in a porous material without it being evident as visible breakdown. The repeated stresses, due to the several phenomena described above, and due to shock, excessive loading, thermal movement and other disruptive forces, can induce in the vitreous or crystalline elements minor failures which are cumulative and can lead progressively to weakness and ultimately to failure. In some climates cycles of freezing and thawing can occur several times a day. In others hydration and dehydration can occur as many as 10 or 12 times. Solar movement, shadows crossing a surface and varying wind patterns may all control the advance and retreat of the frontier of evaporation of internal moisture, and with each movement there comes dilation or contraction due to water intake or loss and perhaps also the expansion of crystals in formation. These repeated stresses move apart the components of the soil structure, leaving a friable material which does not cohere again. The wetting and drying caused by rain or other precipitation simply extend the problem, which is usually most evident at the head and base of a wall.

3.7 The interaction of materials

In buildings raw earths very rarely exist in isolation. They are normally used for walling and are essentially found in combination with

Figure 3.3 Gurayat, Saudi Arabia. Differential weathering in a cement-reinforced render probably due to varying insolation. Render on east- and west-facing walls in tropical zones can suffer higher heat gain due to direct sunlight than north- or south-facing walls on which the angle of incidence is less.

mortars which are likewise porous materials of generally similar characteristics – that is to say, similar porosity, similar density and cohesive strength. Sometimes burned brick is bedded in an earth mortar and earth bricks are bonded with similar earths or with cementitious materials. Lime bricks and blocks are used with earth and lime or cement mortars, while crushed brick has been extensively used in the formation of lime mortars. Bricks and earths are frequently faced with renderings composed of sand, limes and cements, both internally and externally. Stones occur in combination as facing and as integral material, so that the process of conservation must involve treatment of the entire structural mass. These materials are compatible in that they are physically inert, porous, of similar densities, similar compressive strengths and of similarly moderate tensile strengths. Their thermal capacities are similar. These behavioural characteristics group them together. Their relationship to other common

building materials – metal, wood and glass – crosses a divide which provides a convenient definition of a boundary but must not obscure the interaction of the materials when used in combination. In historic buildings an intimate relationship between the porous building materials and glass or metal is rare and poses few practical problems, but wood has consistently been used and misused as a component of porous, inert structures with consequent problems.

3.8 Associated materials

There are a number of peripheral processes which particularly have a bearing on the conservation of earth structures and relate to siting, micro-climate and micro-environment. Structures, vegetation and drainage conditions in the locality can have an important bearing on the conservation of earth buildings.

Figure 3.4 Mixed brick and earth construction. Frost action has caused the failure of an underburned brick without similar loss in the stabilized earth.

Inert materials associated with earths, often at random, in nodular or lump form are stone in every conceivable variety and shape and burned brick. The technology of stone and brick and their conservation is dealt with elsewhere in this series. The porosity and chemistry of stones intimately associated with earths, being closely comparable, rarely induce decay or breakdown that can be categorized separately. Generally there is no deleterious effect resulting from the chemistry of the admixture of stone and earths. Where calcium carbonate is dissolved out of limestones and introduced to an earth mortar, the effect is generally beneficial and it is only with the rare presence of materials such as metallic ores (ferruginous or cuprous) that noticeable reactions can occur. More important is the variation in physical characteristics where a friable or weak stone can fail along its bed, perhaps due to frost action, so inducing a major fissure in an otherwise coherent mass.

Stone and brick are occasionally introduced to give greater coherence or structural strength to walls or vaults. The reinforcement of quoins, the construction of arches and vaults and the introduction of long horizontal courses are particular key features. In some areas, such as south-west France, alternating courses of stone, rubble and earth block have been used widely. In many structures plinth courses, foundations and capping courses are of stone or brick. Brick in particular is used extensively in earth structures for the provision of quoins, ribs, arches and vaults. Much depends on the nature of the mortar used, but even with the softest of setting mortars the effect is to produce a structure very much more rigid than the general calibre of the earth construction, however well-compacted. Only in the situation where earth itself is used as the mortar is there a full coherence in the structure. In the case of rigid mortars being used to bind a feature of brick or stone, the effect is to produce a hard coherent mass, set into a softer matrix. Any adverse interaction is generally physical rather than chemical. Settlements easily absorbed by the more tolerant earth structures may produce uneven loadings upon the masonry, inducing cracking. Arches producing a lateral thrust may receive insufficient support from earth structures, moving much more readily under changing regimes of moisture

and uneven settlements tolerable by the earths, but consequently producing a pattern of failure in vaults. Such failures will normally be consequent upon uneven loadings and more probably the consequences of uneven support. Earths are not consistent by their nature and by the processes of construction, and although in normal work there is a general uniformity, minor inconsistencies run through even the best of earth construction. These are sufficient to allow the development of imbalances in their capacity to offer support. In addition the inequalities in loading due to the differing masses of rigid structures placed on earth construction themselves produce unequal compaction. The failures that occur will tend, therefore, to be those resulting from different resistances to downward movements rather than lateral stresses. The nature of cracking and the strength, direction and persistence of movements determinable by strain gauges or telltales will generally suffice to indicate the areas of inconsistency and weakness.

Organic materials associated with earth structures are either fibrous or structural timber.

Many structural organic fibres have been and are used in conjunction with earths. Of these, straw is the most common and the category can be extended to include all forms of grass and reed, including shredded bamboos. Animal hair is known, although more common in plaster. These materials are used essentially in two ways: as randomly placed fibres which may acquire directional coherence by the method of their working and as deliberately aligned fibrous material. Randomly placed fibres may be macerated or chopped and incorporated into earths in a wet state. These are puddled or pounded in, having been spread on the prepared material and the process generally involves turning-in, wetting and repeating the process until a reasonably uniform mixture is obtained. Present evidence suggests that unchopped straw is stronger in cob than chopped material.

Turning-in of the mix does not necessarily produce a random arrangement of fibres and the subsequent working into the mass material or the render leaves the fibres in alignments which often reflect the method of laying.

Figure 3.5 Mixed tile and earth construction. A fired clay tile used as a permanent shuttering to a lower vault reveals in its failure the fissuring generated in the lower earth walls, the arched haunches and the tendency of earth structures to break away through vertical shear.

Almost inevitably, materials are set down in layers, even where they are subsequently dried so that an earth brick or a mud render will set with a predominance of fibres tending to lie in or towards one plane. These fibres provide coherence in the structure and, on their eventual failure, a consequent loss of that coherence. Not only do the fibres themselves provide strength along their own alignment, but they are universally constructed so that they are stronger in tension along their alignment than transversely. Being complex by virtue of their formation as plant growth, their consistency as a material in terms of both strength and ageing is uncertain.

Some fibres are laid with a designed and purposeful alignment. These include woven matting – reed, rattan, straw – and similar materials laid down in the matrix but deliberately not randomized. Reed and strip bamboo as well as grass and straw may be incorporated layer by layer with mud to reinforce bricks, panels or renders, providing a coherence in two dimensions only, rather than the three dimensions aimed at with random fibre. Analysis of the construction will indicate whether two-dimensional or three-dimensional reinforcement has been provided and careful analysis of the physical state of the material will determine what residual strength this reinforcement will have. Sampling may be done on test cubes cut from the historic material. A count of the number, type and shortest diameter of fibres cut through per square centimetre on each plane will give a comparative measure of alignment, density and size. This information, coupled with an analysis of type and condition, will determine the amount of reinforcement used and surviving usefully.

The breakdown of embedded fibres is generally a consequence of bacterial and fungal attack rather than any process of ageing. Fibre-reinforced earth structures which have remained in permanently arid conditions may have retained their original strengths, whereas those in conditions which encourage fungal and bacterial growth may have lost the reinforcing material largely or entirely. Insect attack normally accompanies the breakdown of cellulose fibres and lignin which will occur in the presence of moisture. Suitable conditions for such decay may have existed at some long time past, leaving the earth structure physically devoid of the original reinforcing material without apparent external change.

One particularly complex arrangement of fibrous reinforcement derives from the natural spread of roots into earths. By cutting turves and applying them to buildings, a complex reinforced fabric of earth and fibres of considerable tenacity has been made available for structural purposes. Turves are built into walls and carried over structural armatures to provide roofs. The root systems of the plants originally growing in the turves provide a very complex interlacement of reinforcing fibres, often in high proportion. From the point of view of construction those turves being more than one-third organic material move into a special category of building material, ceasing to be essentially building earths. The decay of the organic material in this proportion causes irremediable loss of strength, the bulk of the

Although used for millennia, the careful analysis of earth/straw mixtures has been undertaken only recently. Shrinkage and compressive strengths are shown here for one particular Devon soil on the basis of varying degrees of initial moisture content, for shrinkage and for strengths of varying moisture contents.

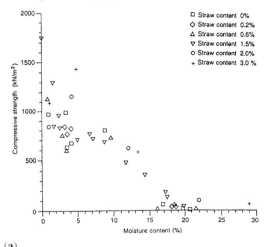

(a)

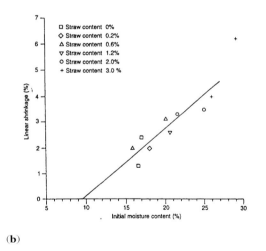

(b)

Figure 3.6 (a) Compressive strengths and (b) drying shrinkages of specific earth/straw mixtures.

root systems vanishing with the decay being carried off as liquids or as food by parasitic life in aerobic conditions. Unless kept dry the material therefore becomes short-lived.

Building structures which depend on the strength of the materials of this type have no permanence. Apart from their use as mass walling, highly fibrous turves have served primitive builders extensively for roofing. In a young and green condition such materials can be very effective, but their shrinkage with loss of water and with decay means that as self-supporting structures their dimensions change quite rapidly. As coverings over other forms of support – principally light timber armatures or weavings – they need constant renewal. Fibrous turves with a high soil content may stabilize as permanent earth structures even after the decay of the root systems. If the admixture of sands, silts and clays is not dominated by smaller particles a stable and permanent structure can result, relying initially on the cohesion by the root fibres. Except in the most arid conditions or in anaerobic circumstances (which in buildings usually means a water-logged below-ground state) the fibres will disappear and the resulting structure will consequently contain a high proportion of voids. This in turn weakens it and makes it more susceptible to erosion. Walls made of turves, in which the upper layer of each contained a higher proportion of fibres, become striated by the effects of such erosion themselves as fibrous decay sets in. Turves are almost invariably laid flat in walling and the results of this type of construction (usually found in archaeological digs) are distinctive. In Ireland and Scotland turf (peat) has been used as walling, laid flat or outwardly inclined on the wall. The Icelandic tradition is to use turf laid herringbone fashion. This ephemeral form of structure is only conserved as an archaeological or museum operation.

Much building involving soils has relied upon a powerful armature as a structural base. At its heaviest this will take the form of a timber palisade, often embedded in the soil (post-fast) or as a heavy structural framing carrying lighter timbers within it. In all such structures the vertical members carry the dead load and horizontal members must be provided to distribute these loads and give the structure cohesion. Such parallel members can range from large posts forming a palisade with excess structural strength to close-set lighter posts, none of which individually would be sufficiently strong to provide the required strength to the wall as

in the mud and stud construction of Lincolnshire – a technique found worldwide.

The strength of embedded timbers on a permanent basis depends upon their nature and the circumstances of the surrounding material. Sapwoods are much more susceptible to decay and beetle attack than heart woods and the highest rates of decay occur in the warm and humid conditions where oxygen is freely available. A timber embedded in water-logged ground may, depending upon its type, survive for a very long time indeed. Its upper section, being alternately wetted and dried, may decay quite rapidly and a higher section incorporated in a building structure and maintained in dry condition may fare rather better, although subject to the attack of larger insects. Such a timber may, therefore, be in perfect preservation, partial decay and total decay over different sections of its length – a common phenomenon in embedded timbers.

Where sapwood fails and heartwood survives a discontinuity will occur between an outer mud structure and its timber armature. Like a turf from which the roots have vanished leaving a series of tubular voids, an apparently massive earth wall may contain a pattern of complete or partial voids as the result of differential decay, the lost sapwood depriving the structure of continuity, although the heartwood remains intact and viable.

While timber of one particular type may decay differentially depending upon its conditions of embedment and exposure, a parallel situation may arise through the incorporation of different types of timber in a historic structure. Many historic buildings worldwide have been built with techniques of timber framing which require secondary armatures within the spaces left by the frame. In English construction the most significant method of this type is the wattle panel incorporated in hardwood framing to provide a base for an earth daub. This will usually be applied in considerable thickness to both sides of the panel with the result that the wattle is completely hidden.

In the poorer examples where coppice sapling has been used the relatively rapid decay of the soft green timbers will leave complete or partial voids. In a relatively small panel where structural loads are carried on the framing this may be both unseen and unimportant in normal circumstances.

Figure 3.7 Parsons Farm, Sussex. A part-destroyed post-medieval wattle and daub wall showing economy in the spacing of staves, and failure in the wattle due to insect attack.

Significant structural loads are taken by the major framing and it will only be in the special conditions of wind pressure impact or other changing stress that the inherent weakness caused by the loss of internal tensile strength becomes noticeable and is treated as a matter of importance.

3.9 Invisible additives

Some organic material sometimes incorporated in building earths is invisible in its ultimate application. Soluble and semi-soluble organic materials such as dung, plant juices, milk and occasionally blood have been incorporated in mixes with advantages that vary with the nature of the mix itself. Where there is a chalk or limestone content these additions are ineffective and virtually unknown. Their useful effect is to plasticize soils with a low or aggressive clay content or to smooth soils with flocculating clays which tend towards excessive expansion. The object of the addition is always an improvement in workability, perhaps simply as an improvement in the speed with which plasticity is obtained. Practical experience of conditions and local circumstances, tradition, folk lore and superstition have all played their part in determining the recipes which undoubtedly can lead to significant improvements in workability and, therefore, density of the finished material. This can be particularly important in applications such as floor finishes and panel surfaces.

In the long term, however, the general conclusion must be that such materials have little or no effect beyond the originally improved density achieved in application. The principal advantage will have been gained from the effects of the albuminous contents which tend to be slimy and, therefore, help to overcome friction between the inorganic components of the earths, allowing them to move more freely and to achieve a better placement. These albuminous materials are eventually broken up by bacterial and fungal action and although the consequent products have no apparently deleterious effects, their removal from the earths leaves the compacted materials slightly weaker.

4

Agencies of failure and their identification

Those same mechanisms of decay which affect other building materials operate on earths but the effects generally are very much faster. The conservator of earths will be conscious that decay in the structures rarely lends them the advantages of a weathered patina: it simply gives the appearance of decrepitude. Earth structures, like painted buildings, rarely gain from the visible effects of processes which can make other types of building venerable.

The agency of damage too often ignored is thermal movement and, although it might be thought that the inherent softness and pliability of earth structures might render them immune to such problems, this is not the case. A study of long earth walls will reveal a tendency to crack at regular intervals, often in the order of 10 metres, though sometimes two or three times as frequently. Vertical cracks accentuated to fissures occur regularly at such intervals and are attributable to expansion and contraction. The extent to which temperature variation is responsible by comparison with changing humidity has not been determined but this type of damage is the product of the annual cyclic change produced by both thermal and humidity variations. Such cracking will also occur at junctions and changes of direction in walls. It is found as vertical cracks running through the structure usually with diminishing effect towards ground level. This phenomenon is not normally the cause of angled or horizontally aligned cracks.

4.1 Water penetration

The effects of water penetration are multiple. In addition to the volumetric changes induced as moisture content varies, liquid water is directly damaging by penetration and by absorption. Where rain or snow is to be expected a covering or sheathing is normally applied to an earth structure. Where this covering fails water will frequently detach it from the surface which it should be protecting, removing it perhaps as a complete skin or slab, or simply by particle decay. A mud render will be eroded, a limewash will be eroded and dissolved, a lime or mortar render will break away in crazed elements, a tar or paint will peel away as a thin flexible skin, facings such as brick and tile will detach themselves as complete elements. Where organic growth is promoted by the presence of water the new growths will actively penetrate the earth. Much remedial work is essentially maintenance by repair and replacement often involving replacement of sections of the earthen substrate. Setting aside the problems of deposition of salts, failure of the outer skin is usually a consequence of incompatibility between the structural earth and the applied surface. Different strengths, elasticities and responses to thermal change can, without the problems of changing water content, give rise to failure of the bond between the surfaces. This in itself need not necessarily be a total disadvantage. Many surfacing materials can survive and be effective as an isolated material, but the outer skin will normally depend upon the core for its structural restraint and, although some freedom of movement between the skin and the core can be helpful, the outer skin will be weakened by loss of support if the materials are in any way incompatible. Any tendency to fracture will immediately admit water which will be

Figure 4.1   Giau Guan, West China. A major earth wall rerendered. The incomplete section shows cracks at separations of the lifts and vertical fissures due to thermal and humidity variations expressed within each lift.

Figure 4.2   A section of the Great Wall of China, built of mud brick in an earth mortar and faced with a mud render. This demonstrates three types of failure: erosion due to run-off from the central tower, aeolian erosion of the base where saline crystallization has made the soil friable, and vertical fissuring (arrowed) (courtesy of Yang Lin, State Archaeological Bureau, People's Republic of China).

Figure 4.3 A semi-circular tower showing the vertical fissure on the centre of the curved face which results from structural movement in opposed directions.

Figure 4.4 A tower at Akda, Iran, exhibiting the common failure of vertical fracture due to the conjunction of differing directional movements.

Figure 4.5 The Great Wall of China in Chinese Turkestan. A sector showing regular vertical cracking in the layered block construction with the subsequent development of runnels in the fissures.

retained with damaging consequence. A further progressively damaging phenomenon is the tendency of loose material to gather at the lowest point of the void of detachment forming a zone of packing which, by rearranging itself, puts pressure on the outer skin with every movement following expansion of the core material – the 'thin end of the wedge' effect. While such problems primarily relate to skins such as render which are stronger than the earthen core they can also arise where the render is of mud heavily laced with organic fibre such as digested grass or straw which forms a coherent mat across the wall surface behaving differently from the core.

Water penetration from the head of a wall or through a roof is quickly damaging to an earth structure and only in arid climates is a mud capping an adequate protection. Elsewhere a wide-eaved roof of a durable and waterproof material is essential to the maintenance of earth structures. The rate of erosion

of earth is generally related to the degree of wetting. Excessive wetting is the cause of the typical erosion patterns in which wider zones narrow the channel into deeper clefts which become thinner and die out as they progress downwards. The diminution of these cracks is due to the capacity of the wall to absorb the descending moisture and is influenced by the tendency of wetted clays on the surface to expand, closing the pores in the earth and becoming temporarily resistant to liquid water. Thus the erosion generally experienced in a wall accepting moisture at the head produces a saw-toothed effect laterally observed as serrations. Vertical structural cracking will attract and offer a channel for liquid water, becoming more heavily eroded than other serrations. The removal of the particles themselves is often attributable to wind action as, on drying out, the friable surface has lost its coherence and is easily disturbed by air movement. The effect

Figure 4.6 Santo Domingo: water entry through cracking of a hard cementitious render followed by lack of maintenance allows plant growth to become established in the earth block structure. Larger plants will quickly exploit water entry into earths with dramatic effect.

ultimately is to leave the structure as a series of pinnacles standing between zones of greater erosion.

The effects of moisture entering the wall from below are exactly the converse. The wall becomes undercut by a coherent cavity developed above ground level due either to saturation of the surface at that level or, commonly in hotter climates, to the deposition of salts in the surface layer. As ground water rises it is drawn outwards and the evaporating water leaves behind it ions which build up the crystalline salt structures, breaking apart the coherence of the soils which are then readily eroded by wind action. As decayed material falls out of the cavity formed so the process continues upwards and inwards, ultimately undercutting the wall structure until it cannot support the weight of the material above. Collapse follows with a

Figure 4.7 Faraj, Iran: the formation of runnels on a vertical surface diminishing due to absorption and creating a saw-tooth effect. This form of erosion is common in circumstances where dry soils are subject to occasional heavy rain. At the impact point the liquid limit is exceeded but absorption reduces water content as drier parts of the wall are reached.

Figure 4.8 Failure of bond between similar material surfaces due to water entry between them.

Figure 4.9 Rainwater discharge from a parapet causing erosion of surface protection and mortar. Failure of a weathering detail will cause rapid decay in soft substrata.

Figure 4.10 Formation of runnel on the alignment of a fissure. Water collecting on the platform above is drained into the channel, accentuating the initial failure. Fissures may be the cause of runnels or may develop within the weaknesses generated by runnels.

series of concavities occuring generally at
regular intervals. While an entire length of
wall may be undercut an unmaintained wall
may accumulate along its base a sufficient
scree of detritus to protect it against the
effects of low-level erosion, but where winds
remove the debris the effect of the undercutting may be predominant and sufficient to
destroy the wall by overthrow rather than by
gradual erosion from the top.

The freezing–thawing cycle is generally less
significant in earth structures than in burned
brick or stone. Although earth building is
common in a continental climate where
winters can produce very low temperatures,
humidity is generally low and it is only in
circumstances approaching saturation that the
formation of ice crystals is important.
Saturation is catastrophic to earth structures
regardless of freezing, so most earth structures
are maintained in conditions where the
moisture content is such as to allow sufficient
porosity to accommodate the formation of ice
without destroying the particle arrangement.
However, surface erosion can be produced by
freezing which occurs just below a saturated
surface in unusual weather conditions. Ice
crystals may form as long stems carrying the
surface outwards to be deposited as a slurry
when the ice thaws. This is a typical decay
process in abandoned structures in temperate
zones.

4.2 Plant growth

Plant growth may be considered at two levels
of size – the larger includes plants with
substantial root systems while the smaller
includes plants on a small or microscopic
scale, including algae and fungi which will
exist within the pore structures of the soils.
The larger plants will often take hold in
crevices which are already areas of weakness.

The growth of roots penetrating the soils will
cause decay by root expansion. They may
stress the wall by the forces exerted by the
plant on the root in wind or simply by weight
or expansion. In decay the root systems
provide tunnels containing food which attract
small animals and plants, extending the
damage.

Plants which are small enough not to
present the problems of true root systems
range from mosses and lichens down to fungi
and algae. They can nevertheless penetrate
interparticular spaces in earths and in growing
they swell to disturb the established interlocking structures of crystal, colloid and plate
arrangements of sand, silt and clay. The effect
is generally to weaken the structure layer by
layer. Organic acids produced in their life and
decay cycles change the chemical structure of
the clay particles. The practical effect is to
weaken the cohesion and produce a powder-
ing of the earth structure.

4.3 Human agencies and animals

Humans are the prime destructive agent in the
history of building. The reasons are many and
sometimes curious. Of all curiosities, perhaps
the oddest is the destruction of buildings by
the extraction of saltpetre from walls impregnated with the product of the decomposition
of urine and excrement. Animal stalls, houses
and yard walls were often severely damaged
in the process of scraping or brushing down
the lower surfaces to obtain a commodity that
was not merely valuable: it was the basis of
political power.

With the invention of gunpowder (an
intimate mixture of carbon, sulphur and potassium nitrate), the ingredients – charcoal,
flowers of sulphur and saltpetre – became a
vital military resource without which the
powerbase of the ruler vanished. Venetian
monopolies on eastern supplies of saltpetre
were circumvented in the 15th and 16th
centuries by European factories in which night
soil was spread on damp earth and ashes.
Urea decomposed into ammonia and by
further oxidation into nitric acid. Sodium (and
potassium) ions combined with nitric acid to
produce saltpetre which dried as a white
crystalline powder. In these factories wood ash
was mixed with water to extract potassium
hydroxide which was then converted to the
nitrate. The same route was used to obtain
saltpetre for glass-making. As an adjunct to
this factory process, supplies were obtained by
scraping the walls of earth buildings in which
similar decomposition had taken place.

Figure 4.11 Bokhara, Republic of Uzbekistan. Severe sustained basal erosion on the city walls, causing partial collapse, and showing the construction strata in the upper levels. The lower section of the wall is protected by a scree of detritus which has not been removed by aeolian effects. The gaps are typical of the failure of earth walls where basal erosion is severe. Entire sections are toppled leaving vertical break lines.

Figure 4.12 Wu Wai, China. The compound wall of the Confucian Temple. The failure at the base (lower left) is due to salinity, thereby losing its two-coat render. The adjoining ground supports vigorous plant growth, indicative of its moisture content.

Figure 4.13 An apparently well-maintained building reveals on the flank incipient failure of the hard cementiticious render. Inequalities in surface materials are a hazard on a weak substrate.

Figure 4.14 Near Lanzhou: northern line of the Great Wall of China. The upper sector of the wall, constructed in consolidated layers, supports little vegetation in contrast to the face of the ditch excavated from native soil.

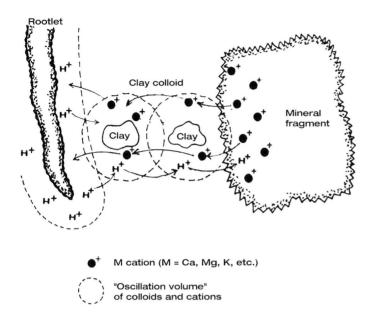

M cation (M = Ca, Mg, K, etc.)

"Oscillation volume"
of colloids and cations

Figure 4.15 Action of roof tips in soil.

The root tip is acidic in itself and additionally may exude carboxylic acids and chelating agents which become acids of the Krebs cycle – tartaric, citric, etc. – by oxidation of piruvic acid. Etching of minerals by the transfer of cations will occur through the mechanism illustrated, with mineral change; the loss, for instance of potassium from biotite produces vermiculite.

The effect is particle disturbance with crystal destruction followed by root pressure through absorption which leads to swelling with damaging results. Resulting localized pressures can be as great as 30 atm (or 3040 kN/m^2). This process can only take place in the presence of free water, as it is dependent on osmotic pressures.

Decaying dung within produced saltpetre outside and 6–8 m^3 of impregnated soil from the outer wall, washed, might yield 1 kg of the valuable product. In the late 17th, 18th and 19th centuries, France required over 8 000 000 kg per annum. It was the stuff and substance of wars. Monarchs protected and encouraged the gatherers, who brushed and scraped the walls of house and byre, sometimes coming to blows with the inhabitants and leaving behind a trail of damage that toppled many a good earth building.

Animals also contribute actively to the decay of earth structures but of all erosive agencies they are the most easily controlled. Larger and smaller animals, including insects, readily burrow into earths. The larger animals are relatively easily dealt with and the damage on this account is usually a consequence of neglect. Smaller animals leave less evidence of their depredations but can be more damaging. Perhaps the most pernicious is the termite or white ant which will create multiple burrows through a wall structure feeding in the earths on the organic content which can be as high as 5% by weight when the initial construction takes place. There are over 2000 species of

termite, most of which live the majority of their life cycle within wood or earth, though some have a flying phase. Their activities will rarely take place in isolation and many other small insects will normally be found in a structure where they are the prime infestation. Structures with earth mortars are particularly badly weakened by this type of attack. An apparently sound masonry structure of fired brick may prove to be pointed up externally with a lime mortar but to be built with earth mortars. A major attack by termites will readily destroy the coherence of the earth mortars and the presence of successive colonies of other insects in addition to the termites can result in the core material being carried away particle by particle, leaving brickwork loosely bedded in dust. Slight settlement of historic walls in an embassy building in the Middle East proved to be of this type, with major security implications.

Both termites and rot can destroy the integrity of embedded timber with immediate direct destruction or loss of bearing capacity. Other dangers must include physical impact of animals and machinery and the effects of vibration from mechanical sources or from

Figure 4.16 Kao Chang, sixth century AD: China. Human erosion. Tourists viewing earth structures by climbing to the highest points of the walling have accelerated the rate of loss of structure.

Figure 4.17 Physical destruction due to external forces such as earthquake, explosion and demolition may leave structures partially intact and repairable.

other parts of the structure in response to wind pressure and buffeting. In assessing weaknesses in an earth structure the mechanical effects of wind on rigid over-roofs and similar fluctuating loadings must always be seen as a possible cause of weakness. High point loads may be derived from relatively modest forces transmitted through rigid materials. By minor deformation a load easily sustained when dispersed may become destructively unsustainable when focused upon a small area. Storm, flooding, tempest and earthquake are forces of major destruction not peculiar to earth buildings, but their effects can be dramatically different when applied to earths whose elasticity and subsequent powers of recovery are minimal and whose solidity is dramatically altered by the presence of water. The remedies and precautions, however, can be very specific because the behaviour of earth structures is very different to that of combustible or more rigid materials.

4.4 Winds

Air movement carries with it certain important effects. It will extract water from a structure by evaporation. An earth structure will have a moisture content rising as high as 20% by weight depending on the dampness of the climate. This figure excludes combined water. Rarely will it fall below 3–4% of free moisture. Consequently variations may range over some 15% of weight. As expansion and shrinkage are a direct consequence of the contained free water (in earths above the shrinkage limit), the effects of evaporation can significantly affect the behaviour of the structure. Rates of airflow will vary according to the shape of the building and the height of the structure. The air laminae closest to the earth will be retarded by friction and obstruction and will move more slowly, producing relatively still pockets of air near ground level, although the upper part of the structure may be well-ventilated.

Figure 4.18 Disruptive forces such as settlement and seismic shock induce fractures which, if not repaired, are extended by erosion.

Figure 4.19 Jia Ohe, western China. The core of a giant Buddhist stupa originally faced with statuary. The lower sections are of native earth, which is less well-compacted and less durable than the same material reconstituted into block for the upper structure.

The shape, size, siting, orientation and adjacent circumstances will affect the typical airflows around the structure, which will have adapted to prevailing conditions. Changed circumstances will induce changes in moisture content with consequent changes in volume, stress, rigidity and structural behaviour. A combination of structures which produces a higher ground-level airflow than previously may induce abnormal drying or wetting or provide eddying effects and consequent erosion. At even moderate speeds air-borne particles can initiate erosive effects and carry them forward as pockets are formed in which eddy currents attain air speeds several times the velocity of the passing air stream.

Small-scale eddying and local vortices can produce air speeds sufficient to pick up loose ground-lying particles and even apparently benign conditions may permit a continuing process of erosion. Diurnal winds occurring regularly as temperatures rise and subsequently fall present an unremitting hazard of relatively small proportions but of significance because of its regularity. The introduction of a new structure close to a historic earth structure can, by changing the pattern of air movement, affect the adjacent wind conditions and, therefore, affect the life and durability of the buildings and its materials. The corner of a building, once stable and durable in a protected circumstance, may be subject (by the exposure caused by the removal of an adjacent structure) to regular drying and airflows, which may be sufficient to carry sand arising perhaps every afternoon due to the effect of the sun's heat on nearby desert areas, commonly producing dust devils. The drying may cause shrinkage due to moisture loss and consequent failure of a surface coating. Its

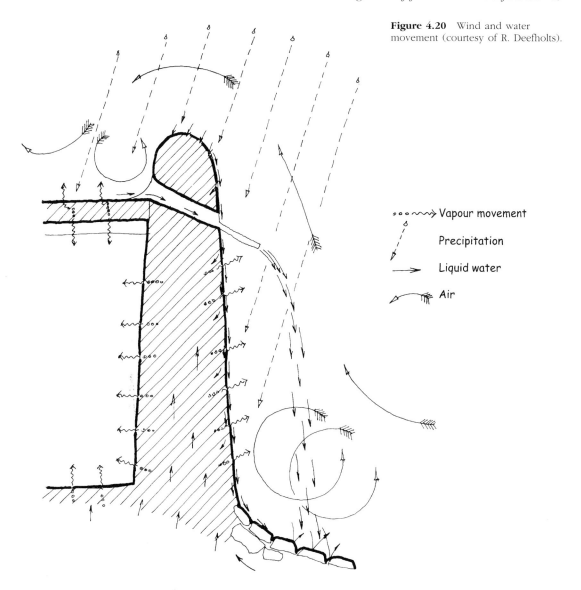

Figure 4.20 Wind and water movement (courtesy of R. Deefholts).

∘∘∘~~> Vapour movement

Precipitation

⟶ Liquid water

Air

removal will be followed by excessive evaporation producing the familiar effect of the
deposition of soluble salts in the exposed
layers of the earth. These soluble salts break
the inherent bonds between the soil particles.
In such circumstances the effect can be to
produce a hard, brittle surface layer on a
weaker underbody with a compensatory loss
of adhesion. It may also cause irreversible
swelling by crystal-expansion between particles, stressing a wall and causing it to topple.

Wind movement lifts up sands and fine
earth particles. These, carried even at relatively
low air speeds on to the surface, cause the
salt-impregnated surface to drop away. The
process is then repeated interminably and
within a relatively short time the base of the
wall may have been undercut and the falling
detritus is carried away by the higher-velocity
winds or well-intentioned street sweeping.

The erosive effects of wind-borne particles increase even in excess of the square of wind speed because, as air velocities increase, heavier particles can be carried. The erosive effect is a product of the size of particle and speed of delivery. An air mass approaching an obstruction enters a zone of increasing pressure which provides the force to deflect it. A solid particle carried within the air stream, having a greater mass and hence more momentum than its equivalent volume of air, is not deflected by the relatively small air pressures and continues a near-straight-line trajectory, impacting upon the solid obstruction at little-reduced speed. The heavier the particle, the less will be the effect upon its trajectory of the divergent air pressures. This effect is coupled with the increased lifting effect of higher wind speeds to increase the destructive power of higher-speed airflows.

The lifting of particles from positions of rest at ground level is due to a combination of lateral movement in the particles caused by friction with the air, or impedance, the reaction of the particles with the solid ground causing them to bounce, and the Venturi effect by which a moving stream of fluid or gas exerts a reduced lateral pressure proportional to its speed. As air speed increases these forces are sufficient to lift and carry along increasingly heavy particles which in turn are delivered with increasing effect upon an obstructing earth wall. It follows that particles of a given size will be lifted when each specific air speed is reached. The result is that with every increase in speed particles of larger size and mass are added to the stream. Conversely, as the air speed falls these same particles drop out to form a loose mass with the detritus produced by their impacts.

Figure 4.21 Due to the curved shape of this ice-house in Kerman, Iran, air speeds around are sufficient to remove much of the material weakened by salinity and crystallization near ground level. Linear walls nearby are less affected.

Since ground-level air speed and direction are partly determined by obstructions, the loss of a structure entirely dissociated from the historic building may well have a directly adverse impact on that building simply due to a change in air movement. Tree growth is another similar factor. Not only does the root structure of a tree change the pattern of available water in the subsoil, perhaps reducing it to the point where significant shrinkage affects the foundations of the historic building, but the interference with airflows consequent on the growth of the tree can dramatically change the exposure and air movement patterns around the building. The interference of airflow produced by a tree rising into the faster-moving air streams above ground level can increase air speeds at lower levels in specific areas, inducing eddying and unbalancing the evaporation experienced by a building. In situations of relatively low water availability the evaporation from the ground surface can be as significantly damaging as the subtraction of moisture from the soil itself and, if the effect is greatly to reduce the moisture available in one part of an earth structure by comparison with another, differential movement can be significant.

The introduction of excessive moisture by air movement can be equally damaging. A structure initially protected from airflow and then subsequently exposed may quickly show symptoms of distress due to a higher rate of introduction of water, perhaps carried as an early-morning condensate on a relatively slow air stream, perhaps merely being a redirected flow of rainwater from eaves or free water carried as rain or snow on to and into the structure. In some instances an unprotected face of an earth wall may have been exposed

Figure 4.22 Severe basal decay accentuated by wind erosion which removes dry detritus, which otherwise would physically protect the damaged layer while causing the erosion zone to rise further.

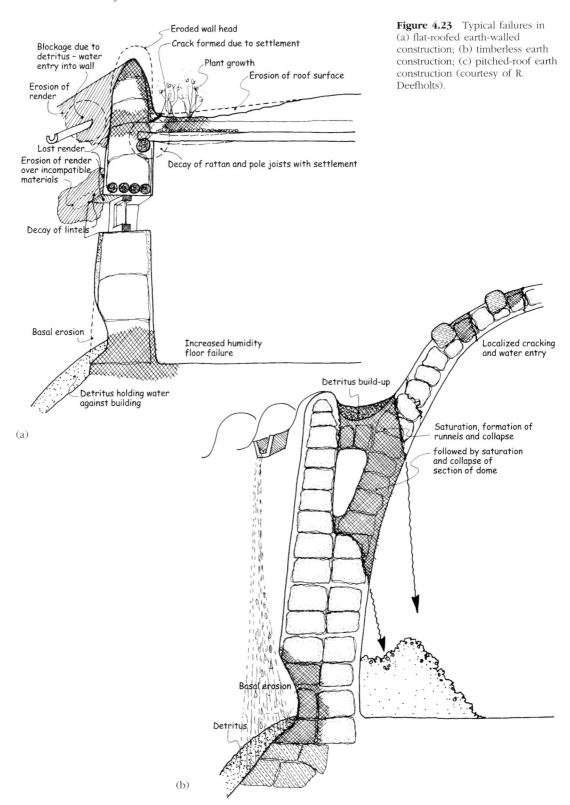

Figure 4.23 Typical failures in (a) flat-roofed earth-walled construction; (b) timberless earth construction; (c) pitched-roof earth construction (courtesy of R. Deefholts).

Figure 4.23 Continued

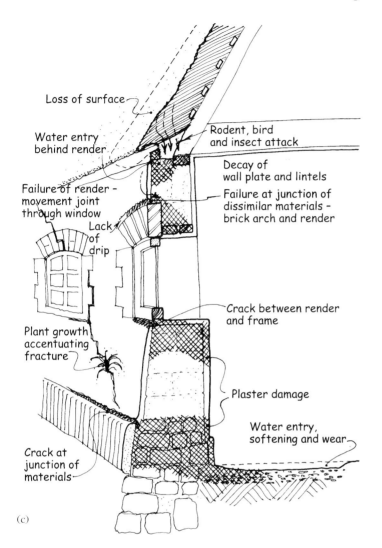

Loss of surface

Water entry behind render

Failure of render – movement joint through window

Lack of drip

Plant growth accentuating fracture

Crack at junction of materials

Rodent, bird and insect attack

Decay of wall plate and lintels

Failure at junction of dissimilar materials – brick arch and render

Crack between render and frame

Plaster damage

Water entry, softening and wear

(c)

by the removal of an adjacent structure, the other walls being weather-proofed. The imbalance, coupled with aggressive weathering, can quickly prove disastrous. In climates where precipitation is periodic this danger can produce particularly significant effects. Slow imperceptible changes, however, can also be very significant. An increased rate of introduction of moisture can allow the build-up of algae, fungi and plant life, changing the character at a rate not generally perceptible but producing significant decay within a relatively short time. Removal of protective structures can have the direct effect of allow-ing winds carrying rain, hail and snow to impact severely on once-protected areas, the effect being to introduce moisture much more rapidly and into parts of the structure not previously affected. These same higher-speed airflows when not laden with water induce a high rate of evaporation, producing in the building a more rapid cycle of expansion and contraction, perhaps exceeding the natural tolerance of the materials, whereas previously in a more protected condition this limitation was not exceeded. A critical area of tolerance in this regard may be the connection between two differing materials, an embedded timber

Figure 4.24 A lack of tensile and shear strength allows cracks to develop vertically with wind and movement pressures acting differentially on adjoining façades. Shear cracks are evident in the walls adjacent to the higher structures.

and its matrix or a surface coating and the substrate. Between these materials there will always be a critical relationship, the bond defining by its strength the level of tolerance which, once exceeded, causes the breakdown of the relationship.

Thus, two materials with similar coefficients of thermal expansion and an adequate physical bond might be expected to be compatible.

However, if the coefficients of expansion due to humidity are at variance with those for thermal expansion, then changes in the available free water will produce differential expansion and contraction regardless of temperature. Water may be supplied through the air being absorbed directly into the pores or simply as vapour producing a condensate, or directly as precipitation. Other water may

Figure 4.25 Water delivered where not intended causing localized failures. Failure of downspouts (or gargoyles) intended to throw roof water clear of a wall; initial failure develops into vertical erosion through lack of maintenance.

arrive second-hand through precipitation which has fallen elsewhere (on to a roof or the ground) and then has been drawn into or delivered on to the structure. The changing rate of air movement may have carried the relative rates of introduction and removal of water to a point when the coefficients of expansion produce stresses which exceed the cohesive strengths between the substrates and in consequence there is a breakdown. This effect is particularly observable in earth structures (usually in combination with parallel effects) due to the weak bonds between materials.

Circumstances and conditions are so variable that measurement and calculation become exceedingly complex – justifiable in the design of a supersonic airliner, but impractical in a single historic building. Assessments must, therefore, be made on the basis of experience. The fundamental requirement must be to take account of conditions of siting and relationships with adjoining structures. Even more critical can be the introduction of associated structures built perhaps with the intent of conservation. Overroofs and enclosures can have – and are intended to have – a profound effect upon the micro-environment, perhaps protecting visitors, sheltering structures from saturation and the erosive effect of winds and keeping out pests. With every attempt to exclude a danger the probability of a further mechanism of decay being introduced can arise.

4.5 Recording and analysis

Built into the fabric of every historic building lies information about its history and often about its creators. This is contained not only in the physical structure and the methods of its construction but also in its materials, and the conservator will always be conscious that every structure contains information of which the present generation may not be conscious and to which it does not yet have access because the technology is undeveloped or unknown. Like many analytical techniques now available, it may not even have been predicted. Earth structures have a particular importance in this regard. They are essentially natural materials of their time, relatively unaltered, and they often relate to periods beyond written history even when those periods may be relatively recent.

The invention of techniques of recording is accelerating and increasing improvements are providing critical refinements. Physical measurement, probably the oldest technique of scientific recording, has been applied to earth structures relatively late and inefficiently. Their qualities often lie in soft contouring, generous mass and uneven form. Consequently the task of recording is more complex and less mathematical. Two technical advances have provided new opportunities – the speed and accuracy of optical surveying using laser technology and the invention of photogrammetry. Both represent an important step forward in the ability of the conservator to record existing structures. Photogrammetry is particularly valuable because it allows the irregular nature of surface contouring to be picked up and graphically described prior to the interventions that go with restoration. This technique has the secondary advantage of producing a permanent record encapsulating material information which may not be delineated on the drawing but is nevertheless available for retrieval at a later date. The principles of binocular or stereoscopic photography have been applied for over a century but the availability of high-precision equipment and computer-aided analysis now offers greatly improved results. These techniques and other methods of physical recording are the subject of specialist works, but other important techniques lie in geolog-ical and technical analyses. As technology advanced in the last century it became possible to analyse the chemical constituents of soils. The admixture of many different compounds in soils produces a distinctive ratio of components in each sample and this can serve to identify the origins of the materials. Chemical analysis has moved forward with the invention of techniques such as spectroscopy producing swifter and more accurate results and the conservator can now turn to laboratories for soils analysis both in the chemical and in the physical senses.

4.6 New techniques

In the repair of buildings the physical and chemical profiles of the original material and the replacement material can be deliberately varied so that new material can be readily identified. Unless great care is taken in the selection of additional material or the reuse of soils, this differentiation tends to occur by nature of the operations, often due to the introduction of a modern material which acts as a trace. In historical terms there is a strong case to be made for the placement in repairs of trace materials which will identify the date of the work. Such materials can be as straight-forward as the deliberate use of modern adhesives, preservatives or synthetic fibres. The deliberate inclusion of a pattern of trace elements in granular indissoluble form is equally possible.

Geological analysis can perform similar analytical functions and mineralogical laboratory techniques such as high-resolution microscopy and diffraction analyses are capable of identifying the sources of soil components. The nature of the admixtures offers the conservator additional resources for investigating origins and the corroboration of conclusions reached by other methods. While these techniques may be relatively unimportant in normal practice, being unduly sophisticated for those involved in general conservation, they become crucial to the solution of archaeological problems and may well have a bearing on the replacement materials used. Improvements in techniques of microscopy and in particular the availability of the scanning electron microscope have probed the particle structure of earths much more deeply. The resultant analyses are beginning to aid the conservator in understanding the behaviour of synthetic materials introduced for purposes of conservation in addition to providing data on material behaviour at the molecular level.

A further range of techniques introduced relatively recently relates to the analysis of organic material which is usually present in earth structures and can give important data about conditions at the time of building. Analyses of root material and introduced fibres such as straw and hair can identify the circumstances, date and nature of construction, allowing the contemporaneity of structures to be ascertained. A very specific and important level of activity has arisen out of spore and pollen analysis. Evidence of the interrelationship of plant growth and building structures applies to the problems of building history and to those of climate and environment, allowing important linkages to be established and clarifying dates of construction. These analyses depend upon the identification of the wide variety of microscopic particles produced by plants both at pollination and reproduction as spores.

Other microscopic vegetable and animal tissue likewise becomes available through the same techniques through areas of study such as genetic structure, which can be said to have hardly begun. In some cases their availability in a structure of identifiable date will have been encapsulated and maintained in a quasi-permanent condition, providing a bedrock of evidence for enquiry in the natural sciences rather than in building history, although additional information becomes available as a byproduct. An earth structure yielding fossilized insect and pollen material may reveal information about climatic conditions at the time of its building which in turn may contribute to an understanding of the reason for the inclusion of specific features in its design.

A technique close to the popular imagination began with organic matter obtained in the analysis of building. A naturally occurring isotope of carbon, with an atomic weight of 14, is radioactive and therefore decays at an identifiable rate. The process of random nuclear decay changes its valence, reducing it to another isotope of carbon with a lesser atomic weight. As carbon-14 occurs naturally in the atmosphere its introduction into organic material takes place during the lifetime of the plant or animal and ceases upon death. The process of radioactive decay, however, continues diminishing the proportion of carbon-14 which remains in the structure, therefore measuring the distance in time at which death occurred. The technology is exceedingly delicate and is only effective at substantial distances of time. It is generally relevant to earth structures only in an archaeological context and depends on the survival of organic matter. A similar technique of recent development and still in the early stage of development is optical dating (OD). This technique depends upon a distinctive behaviour of quartz, which is the most common constituent of soils. Silicon dioxide in the form of quartz crystals has the particular property of achieving a specific molecular arrangement within its lattice structure under the impact of light. The pressures of photons cause the electrons to adopt a precise arrangement within the lattice. When a quartz crystal is buried in soil this arrangement can be disrupted by natural radiation within the ground. This radiation varies from place to place but is measurable in any particular circumstance. The degree of disruption to the electron arrangement depends upon the extent of interference in the ground by natural radiation and the longer the crystal remains there, the greater the disruption. Effectively the electrons are dislodged by the absorption energy from the natural radiation, storing the absorbed energy as potential energy. On re-exposure to light the electrons resume their original position and release the absorbed energy as electromagnetic radiation (a pulse of light principally) which can be measured. The energy release corresponds to the time that the crystal has spent in the ground after its previous exposure to light and hence provides a method of dating which is similar in principle to the carbon-14 dating system. This extremely delicate technique is still in its infancy. More fully developed dating techniques include dendrochronology, which is applicable where associated timber samples are substantial and where a matching pattern of dated evidence has been established to provide a time frame, while obsidian rehydration is a further technique which similarly depends upon the measurement of a slow process of change.

Processes such as these provide methods of recording and analysis that can allow the history of a structure to be recorded in considerable detail before intervention takes place.

5

Evaluation, preparation, testing and replacement of materials

Earth is a versatile and durable building material. The techniques of building in earth have always been essentially simple and so, until recently, have been the methods of analysis and repair. Vernacular builders have developed techniques which have enabled them readily to assess the suitability of the material and these remain applicable to conservation work, although supplemented by more scientific analysis. The essence of evaluation for construction purposes is to identify the mix of particles and contaminants. By the nature of the material the silicas are larger; micas, chalks and feldspars are medium in size and the clay aggregates are very small. Contaminants are largely humus and salts which relate to the pH factor. The objective of evaluation is to identify the proportions of these components. Supplementary tests are then aimed at identifying the behaviour of the mix, and the nature of the components in each fraction. The extent of incorporation of chalks and gypsum can be important, the types of clay even more so.

5.1 Simple tests

The first and most basic method of understanding the nature of the available materials is feel. The plasticity of an earth depends upon its moisture content and the nature of the materials. In broad terms, the moisture content can be readily assessed by straightforward estimation of dryness and wetness – the product crumbles to powder at one end of the scale and dissolves into an amorphous liquid mass at the other. At a moisture content giving tolerable workability, the proportionate content of large particles can be assessed by grittiness and lumpiness. The residual material, as an extracted sample, will betray the relative proportions of sands, silts and clays by feel when rubbed between the fingers. Dampening off the residues and testing their slipperiness will identify the proportions of clay as opposed to sands in the general terms which will satisfy experienced craftsworkers or conservators. A more sophisticated sensing by feel can be achieved with the teeth. While the particles of clay cannot be sensed in this way, the granular quartz particles in sands in a very small sample are readily, if unpleasantly, detected as gritty. Silts are slightly gritty, and distinctly identifiable. This test is not to be carried out on biologically contaminated material. If in doubt, sterilize by baking.

Visual inspection is not to be ignored. Colour can be informative when the nature of particular soils is known. Clays can vary from blue-grey through grey-greens and yellows to white and into a range of browns and reds as the ferruginous content increases. Differing colours of clays are often associated physically. While some soils such as loess and laterite with uniform consistency or preponderance of one particular structure can be discouragingly uniform, other beds can be very diverse and colour can tell the experienced worker much about the nature of a material.

Circumstantial evidence is often telling. Observe the landscape. Vegetation type is a primary indicator of pH values. The presence of iron oxides is usually evident in clays by red and brown coloration, but clear water from acid soils may also be high in iron. Where chalky deposits or limestone pebbles are much in evidence the soils may contain chlorites, carbonates and alkaline montmorillonite clays.

Montmorillonites are likely to be the clays present in basaltic areas, particularly with poor drainage. In better-drained – more oxygenated – soils montmorillonites will be succeeded by illites, but this is a matter of proportion. Neither clay is found without the other, if only in small proportion. The mixture may be known as smectite.

Shales will tend to produce soils with montmorillonite–illite (smectite) mixtures of clays. Montmorillonites also tend to predominate in saline clays, particularly at depth and perhaps exposed by excavation or erosion. Run-off which is cloudy, brown or yellowish will suggest montmorillonite–illite mixtures in suspension, particularly if the water fails to clear when standing. Montmorillonite predominance in clays may produce greys, dark greys and even charcoal colorations, whereas the predominance of illites offers oranges, browns and muted reds.

Laterites tend towards the deeper reds and are occasionally evident in mixtures as hard nodules or small seams. Iron ores are similar.

Clear run-off may indicate an absence of clays but also, where deposits are mottled white or grey-blue, kaolins and/or bauxite may be present. These colours may be masked by iron deposits where calcium and magnesium are readily available, giving orange or red colorations. Kaolinites are often lower-stratum material found in areas of granitic erosion.

Micas are often seen by the bright reflections from platelets in soils or compacted soils (sedimentary rock or shales).

Carbonates (limestones and chalks) often betray their presence as discrete fragments on close examination, being crumbs of white or grey and soluble in acids.

The cracking and porosity of soils are also useful indicators. Open textured soils with substantial clay content suggest a high proportion of allophanes (which are not powerfully cohesive) and some kaolins. If clays are absent this characteristic suggests carbonates.

Deep, widely spaced cracking (typically 250 mm spacing) as soils dry indicates a predominance of illites. Deep, close cracking (typically 50 mm spacing) indicates the predominance of montmorillonites.

Straightforward visual inspection gives an immediate assessment of particle size in the upper range and this can be enhanced by smearing a sample across a transparent base and observing it closely in transmitted light. This is particularly helpful in gauging the gradation of particle size across the medium and upper range and the use of a simple magnifying glass will enhance the perception. A lens held at a standard distance from the sample can be a valuable tool. A printer's lens is ideal. As in all these tests, the value judgements depend upon experience with the material. A sample of known working characteristics and known performance can be used as guide.

The sense of smell is only useful in determining whether organic constituents are present and whether the material has come from anaerobic conditions. This information is of minor value. The knowledge gained from visual and tactile examination can be enhanced by testing at the ends of the moisture spectrum in addition to the simple tests applicable to workability. This type of analysis goes beyond the tests normally applied by traditional craftsworkers. Samples dried or fully saturated can be analysed in powder form and in dispersed solution. Powders agitated on a slightly sloping paper surface will quickly display particle size in sufficient detail to allow proportionate assessment to be made. Likewise, the residue lying on the bottom of the glass vessel and the suspended solids in aqueous solution allow similar analysis. A practised eye and a good magnifying glass or low-powered microscope are sufficient equipment.

5.2 Field station analyses

An aqueous test (sometimes known as the Emerson test) is used to indicate the type of clay in an inorganic sample. The test should be repeated several times for consistency. A small coherent piece of soil (about $10 \times 10 \times 10$ mm) in dry condition is placed in an excess of distilled water in a transparent vessel. If after a few minutes it has swollen without collapse, it is probably lateritic. If it swells and then collapses and disperses it is smectite, from which the illite and montmorillonite components may be estimated by examining a crumb in distilled water against a dark background. A fine halo of particles is given by montmorillonites. The more

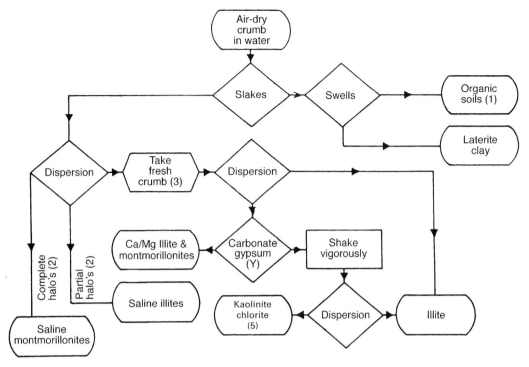

(1) If the crumbs are not air-dry at first immersion, the scheme is still valid, but this category will contain the non-saline illites and montmorillonites as well as organic soils.

(2) Dispersion is most readily detected by the formation of fine misty halo's around each crumb, easily visible against a dark background. The more pronounced the halo, the more dispersive the soil.

(3) Take fresh crumbs and moisten, remould lightly, and immerse,

(4) The presence of carbonate, if not already recognized, can be readily verified by effervescence in the soil when a drop of acid is placed on it.

(5) Settling to a clear supernatant liquid in less than 10 minutes is to be regarded as non-dispersion.

Figure 5.1   Emerson test procedure.

pronounced, the greater the proportion.

Taking a further sample, allow dispersion to take place and test for pH. Acidity indicates a significant amount of organic material in decay and possibly iron salts. Alkalinity indicates hydroxides, chlorites and carbonates.

Taking a further sample, immersing as before and shaking until in suspension will differentiate between smectites on the one hand and kaolinites and chlorites on the other. Smectites remain in dispersion beyond 10–15 minutes.

Follow-up tests include immersion in hydrochloric acid solution, when effervescence indicates the presence of carbonates. Testing

with silver nitrate and separately with barium chloride will in the first case indicate the presence of chlorides and in the second, the presence of sulphates, of which proportions can be determined by titration.

The presence of organic material – but not its volume – may be shown by carbonization. However, a sample immersed in limewater for 48 hours will, after shaking and resettlement, yield a dark colour in solution if the soil is humus-rich.

Neutrality between alkalinity (basic) and acid conditions may be determined by the use of methyl red, phenolphthalein solution or, with some reservations, by litmus, and these materials can give evidence of degrees of alkalinity or acidity. Simple testing devices are available for horticultural purposes.

Physical tests of workability include the preparation of small rolls of the material and their bending to breaking point, and simple tests of extrusion between the fingers in a state of plasticity. Tests of this type reveal general workability but the results vary within a given sample depending on the size and number of larger grains contained within the fracture point or the zone of extrusion. More reliable physical tests of an equally simple nature involve the actual working of samples with trowels or spatulas on a palette at various consistencies with additives as selected. These are perhaps the most fundamentally satisfactory tests of all because they can be coupled to shrinkage tests. Small samples can be worked on impermeable and semi-absorptive backgrounds in comparable areas limited as to depth and size by battens to provide consistent sample material. The physical characteristics of the sample can then be assessed in terms of workability directly related to shrinkage characteristics.

An interesting and direct comparison can be made if the same samples are then applied to a background which does not resist shrinkage movement on drying. Such backgrounds are principally plastic surfaces such as ultra-thin polyethylene and coatings of Teflon on a firm background. The merit of such simple field tests is that they identify the performance of samples physically on the basis of the workability against shrinkage characteristics and the experienced conservator or artisan will assess the comparative performances against the visible admixture of different particle sizes and possibly different types of clay. Experience in evaluation is essential. It will be recognized that the mixture of greatest cohesive capacity is not necessarily the most advantageous material. The more powerful the cohesion, the greater the distance between the cracks which must form due to shrinkage but the wider the crack when formed. Such processes however are applicable to comparative analysis in the field in circumstances where samples of existing material are to be matched with material to be used as a basis for intervention. Testing may conveniently be carried in plastic channels of no more than 20–30 mm cross-

section. The comparative cracking on dehydration is then related to proportionate analyses from which clay content is derived.

Tests of this sort, and other types of performance testing, are essential field work at the design stage of conservation of earth structures. They can be extended to achieve other types of comparison such as weathering, receptivity to finishes, acceptance of plant growth and comparative performance where additives are used to control factors of decay.

Beyond the basic tests that can be carried out anywhere there is a range of practical evaluations which need only the facilities of a simple workbench laboratory such as can be set up in a site hut.

Compressive strength of a simple hardened block can be determined by straightforward loading; dead weight or pressure is superimposed until crushing is achieved in free or constrained situations. The purpose of such tests will usually be to assess the comparative strength of the material to be introduced. The comparison will, therefore, be between new or reconstituted material brought to a density matching the historic material. There would be little purpose in such tests if the two materials were similar in composition but the merit of field-testing is to establish any difference in performance where a new material is introduced as part of a proposed scheme of intervention. Additives and mixture variations are frequently proposed and, where they are proposed with good reason, field-testing may be appropriate.

In carrying out dead-load testing it is important to reflect – but not necessarily simulate – the conditions in which the new material will be used and its purpose and function must, therefore, be understood. For instance, if the material to be introduced into a historic structure is to be in a condition where it will be loaded in a confined space (i.e. with the lateral restraint given by a structural mass), the circumstance can be simulated by loading it within a tube or open-topped box and applying the dead load to a plunger. If the new material will not receive lateral restraint in the structure it may be tested by forming the sample as a cube or cylinder and applying a dead load to the top surface. This unconstrained form of testing is less complex and much more reliable in interpretation.

Figure 5.2 A tower, now an isolated fragment of the Great Wall of China, displaying vertical fissuring which causes the loss of segments in vertical slices. The loss of lateral restraint in the lowest zones effectively 'unloads' the compression previously applied to the bottom strata by the mass of superimposed earth. The lateral stress-release causes the peripheral mass to form shear cracks parallel to the external surface. This phenomenon is unique to earths, among building materials, due to its compressibility and low shear strength.

In any application of dead load the loading member must be restrained so that it moves vertically but cannot move laterally and the angle of application of load cannot be altered. In the simplest terms this means that a known weight should be applied vertically to a test cube. All such tests should be carried out as a series to avoid reliance upon any one sample.

Further field tests can readily be undertaken to determine physical characteristics of samples to evaluate comparative performance, including porosity, density and solubility analysis. These tests again involve making up comparative samples of reconstituted or of proposed new material and these are best done as test cubes formed in a simple mould. Techniques and materials should be those that are used in the field and the samples should, therefore, be made up using hand-mixing methods and trowel application. Operator errors can be minimized if one operator performs each series of tests. The information gained is within the normal parameters of accuracy consistent with general work on site and these may vary quite widely by comparison with the levels to be achieved in laboratories. The conservator must, therefore, make a judgement as to the application of results obtained in field-work by comparison with those obtained under more rigorous scientific conditions.

Practical testing may diverge from scientific understanding and it is important that descriptions are meaningful in terms of conservation. Thus a scientifically accurate description of the percentage contents of metallic elements in a sample may be impeccable but useless to the conservator who has no means of knowing from that analysis what the material is in practical terms. Pure carbon might be soot, graphite or diamond: 95% calcium carbonate might be limestone, marble, lime putty or chalk! To the particle physicist the conservator's sample will be space-containing discrete packets (quanta) expressible in mathematical probabilities. To the geologist the component particles may be identifiable by distant source of origin. To the analyst the pores will be measurable by the extent to which a liquid can find lodging in them. Fortunately a summary of basic analytical procedures has been prepared by ICCROM (Teutonico), in which detailed testing methods are described in the practical terms of a laboratory technician while the general principles of analytical work are explained below. The technical manual should be at the analyst's elbow and should be a guide to the practising conservator.

5.3 Analytical objectives – porosity and rigidity

The conservator requires a number of straightforward physical standards by which to evaluate the material. Porosity, for instance, may be judged by simple visual examination. The application of drops of water of consistent size on a flat surface of a dry sample will illustrate absorptive capacity which is a combination of porosity and the degree of water attraction or repellence (hydrophilic or hydrophobic properties). This absorptive capacity is of practical importance to the conservator who may wish to determine it in the field, particularly in the context of the water repellence or structural rigidity induced by the introduction of materials specifically designed for this purpose. For natural (non-hydrophobic) material a simple test can be devised as follows:

From the original material a small cube (say, 30 × 30 mm) is cut with care or formed in a mould by reconstitution).

A flat slab of porous foam is maintained in a tray in a fully saturated condition and on the foam is placed a weighed, oven-dry sample cube of material under a glass bell or similar impermeable cover. The volume of the cube has been measured physically, or by manufacture of the sample in a mould whose volume is determined by measuring the quantity of liquid required to fill it. The sample is allowed to draw up liquid until saturated, and is then weighed to give basic information on absorptive capacity. For more sophisticated analysis mercury immersion is not now used, as it has been replaced in the laboratory by gaseous displacement. An irregular sample can be volumetrically measured in this way to a high degree of accuracy in laboratory conditions.

As water is absorbed by capillary action, the sample eventually becomes fully saturated, at which stage it collapses under gentle vibration. If it does not collapse the material has been sufficiently consolidated by natural or synthetic processes to be classified as a mudstone or shale, or a consolidated earth.

If the experiment is required to give a precise value of porosity, the foam base must be replaced by a sheet material on to which a minimal depth of water is introduced. The sample then being carefully removed is weighed, heated until dry and re-weighed. The time between placing in position and total collapse is measured. The weight and therefore volume of absorbed water being known, the voids accessible to water can be defined in relation to the volume of earth. If the sample did not collapse it may be saturated by immersion and weighed before being dried and re-weighed. The results of such a test are an absolute value of porosity and a comparative value of absorptive capacity. The time taken to the point of collapse by various samples of identical size but varying material is an indication of their comparative absorptivity.

The weight of water expelled from the material during drying can be converted to volume, which volume can then be expressed as a percentage of the total volume, giving a measure of the available pore space. This measurement may be less than the absolute total porosity by virtue of the retention of some air within the collapsed sample. The difference, however, will be small, being due to the expulsion of air during collapse and collection.

The porosity and the density of any sample depend upon its compaction. A sample formed as a regular geometric shape – a cylinder or a cube, for instance – can be weighed and the density can be expressed against the equivalent value of the same volume of water, which will give a ready method of comparison between samples and with other materials.

Therefore a simple variation on the foregoing test can be based on the use of a standard tube into which a known volume of sample material is compacted using a spring-loaded compactor to control density. Dry weights and saturated weights may then be compared to establish absorptive capacity in the same way.

A range of other tests can be devised to evaluate the physical properties of building earths usually on a comparative basis, in which other samples of known provenance or characteristics are used as controls.

Workability, for instance, can be tested by extrusion. A plug of the material in plastic condition can be forced through a hole in a plate by means of a hydraulic jack operating in a cylinder. The pressure on the jack will give a measure of workability and, more directly, the length of horizontally projecting extrusion carried by the cohesive strength of the column before it breaks away will provide a practical measure of the ductility and cohesion of the material. The comparative results of any such experiment can provide useful guidance where they relate samples of known capacity to those on which information is sought. The objective of such experiments must be to determine within quite broad limits of tolerance the suitability of the mixes to be used in repairs or interventions, with the general objective of understanding and defining compatible physical characteristics. A workability test used widely in the ceramic industry shapes a cylinder of clay in a mould using a piston and dropping weight. The deformation of the cylindrical sample without the mould when impacted by the same piston is a workability index.

Penetrometer tests are of the same general nature. They are now a standard measure by which the rigidity of a malleable sample can be gauged.

There are few ready experiments for determining the behaviour of new materials over a long period of time other than comparative weathering tests, not all of which can be accelerated readily or reliably, however.

An important physical characteristic of earths is the variation in their stability in wet conditions. Slump is a well-recognized characteristic occurring at the point when sufficient water has entered the structure to fill the pores and release the cohesive effects by allowing water pressure within the material. This overcomes the surface tension and the mutual attraction of the particles and plates to each other, allowing them to move apart freely. This over-simplification describes the condition where the earth sample will have lost cohesion. This condition, known as the liquid limit, will occur before the total saturation described above in the absorptivity tests. Its importance is that it defines the point of loss of strength at which the material becomes non-cohesive and non-load-bearing. A sample in a dry condition may be placed under a small weight and fed slowly with the amount of water expected to be needed for saturation. Loss of rigidity and load-bearing capacity occurs before the absorption of the full amount of water needed for saturation. The water yet to be added when this point is reached can be measured by comparison with the total volume needed for saturation and an average over a number of samples can be used to demonstrate the capacity for the mix for stability in wet conditions. Materials containing cementitious components may, of course, remain rigid even when fully saturated but the test, like others, is of value where it is comparative. Acidity and alkalinity may be of interest to the conservator in some conditions, when material matching is at issue. Colour card indicators are now available to allow all soil pH values to be established with sufficient accuracy for all practical purposes.

One other range of tests appropriate to a field laboratory is analytical and these tests probably have greater relevance to the practical work of the conservator than the foregoing.

5.4 Particle size

It is well-established and much repeated that the relationship of particle size by proportion in the mix is the key to determining a successful material and it is therefore important for the conservator to be able to break down readily the components of any mix to determine the particle size of its constituents.

In the most basic form, sieving of the dried powdered material yields immediate information on percentages of specific particle size. The technique is of particular use in the sand/silt range. The clays, however, attaching readily to other larger particles or forming nodules of silt size, are not readily accounted by such methods as running across wire mesh screens and ultrafine sieves. Nevertheless, a set of sieves is a principal piece of apparatus in the field laboratory. A weighed oven-dry sample is crushed in a mortar by kneading under a rubber-covered pestle until all grains are freely moving. The sample is weighed. The mixture runs through the meshes in sequence to determine the percentage weights which pass each successive size, the most useful gradations being coarse sand, fine sand and silt. Comparison of the total finished weights with the original weight will confirm the accuracy of the process. In the event of significant loss the test should be repeated.

In the laboratory, for greater accuracy in the range of finer particles, controlled dispersion in water (with Calgon as a dispersant) allows analysis down to 300 mesh. Below this level sedimentation techniques and X-ray analysis are used. Such results are unlikely to be required by the conservator.

In practical terms the simplest form of particle-size analysis is to put the sample into solution with excess of water and a small amount of detergent. The particles are shaken up in a glass tube and allowed to settle, be suspended or rise over a period of upwards of 12 hours. By this time much, if not all, of the organic material will have formed a layer

at the surface, the bulk of the water will contain the finest clay particles still in suspension, and below this will be layers of clay particles, silts and sands, with the largest components at the bottom. The boundaries of definition will usually be clear except for the gradation of the finest sands into silts.

An assessment of the relative proportion by volume can be made by simply measuring up the side of the glass tube. If the liquid is poured off and the contents of the tube are allowed to dry, careful extraction will enable proportions by weight to be determined with sufficient accuracy for all practical purposes in determining a mix. The principal use of the test is again comparative: a sample of satisfactory material will yield one pattern of sediment and the mix which is sought will yield a pattern which is broadly similar. A more sophisticated version of the same test employs variants on a method of separation by pouring off the liquid containing suspended particles. A popular version makes use of a factor of three in a time sequence. The sample is suspended in solution by being vigorously agitated and then allowed to settle for 3 seconds: the liquid is then drained off. The residue which has settled out is the aggregate and grit component. The material poured off is retained and is again agitated. It is then allowed to stand for 30 seconds and the liquid is again poured off. The remaining deposit is the sand component. The same procedure is adopted and the sample is allowed to stand for 3 minutes. The liquid is again poured off and the residue is the silt component. The liquid is then allowed to stand for 3 days and it is then poured off, the residue being the clay component. At this point the residual liquid contains the soluble salts, some extremely small particles still in suspension and the organic material in flotation. The separated samples can be dried and weighed and the proportionate relationship established by weight, or if required, by volume in a dry condition.

A more sophisticated version of this test can be devised quite simply in the following way.

A channel forming a deep and narrow trough typically 1 metre long, 300 mm deep and 10 mm wide is constructed with glass sides. The trough is water-filled and a continual feed is introduced at one end sufficient to replenish the contents in 3 days. The discharge at the other end passes through a sieve before going to waste. The sample material is introduced in a state of agitation and suspension at the inflow point. This sample is the product of draining off after 3 minutes to eliminate coarse sands and aggregates. The water is treated with a small quantity of surfactant. The introduction may be phased over a period of some hours and is made in conjunction with the inflow so that all suspended particles begin their trajectory at the same point and the entire liquid stream traverses the settling tank in a 3-day period. After sufficient material has been passed through the process a continuous gradation of particles can be seen to have settled along the bottom of the tank in a clear plastic channel or paper-lined trough placed there for the purpose of recovery. After draining, this material can be analysed by particle size and nature. Extremely fine particles and organic material will have continued in the emerging fluid which can be sieved and filtered for analytical purposes. The merit of this test is that it produces a deposit in which every successive particle size is separated in gradation.

Though by no means definitive, this series of tests runs the gamut of useful analyses to be made in a basic (field) laboratory. It will generally provide conservators with all the knowledge essential for understanding the materials of a historic building with the mixes which they will be able to introduce to it. The conservator's work, however, can be backed by a large range of physical and chemical analyses, to say nothing of tests involving biological cultures and botanical analyses, to be made in permanent laboratories. Few of these techniques are specific to earth structures and their relevance will generally be related to testing for specific performance or determination of constituents. Occasionally their importance will be related to interactions between additives and base materials.

5.5 Laboratory analyses

Geological laboratories can provide data on the nature and origin of particles with the objective of identifying sources by the nature of the mixture and identifying the specific clays in the binding fraction. The types of clay can be critical to the performance of earths in some situations and this information may in some circumstances be helpful or even critical to the actions of the conservator. The characteristics of soil components and their performances are summarized here.

Soils with predominant particle sizes in excess of 2 mm may be classified as:

- well-graded gravel if particle sizes are all adequately represented
- poorly graded gravel if one grain fraction is predominant
- clay–gravel mix if smaller particles give cohesion
- silt–gravel mix if smaller particles fail to give cohesion

When more than half the particle sizes are less than 0.08 mm classification may be as follows:

- sandy soil, well or poorly graded depending on the even or uneven distribution of particle sizes
- silty or clayey soil depending on cohesiveness

Electron microscopy is now capable of revealing the structure and performance of materials even at levels approaching molecular size. The results are of secondary interest to the conservator. The information they contain is always subject to interpretation by specialists but it does contribute fundamentally to understanding particle behaviour at the smallest scale. This information can provide the ability to measure change at a microscopic level, to monitor the performance of new and introduced materials and to predict future performance.

Physical and chemical analyses provide identification of salts and other foreign substances in an earth structure. This information may allow the conservator to identify the source of impregnation and assess their permanence, continuance and deleterious effects. Further specialist advice may be essential before conclusions are reached and interventions are made.

Tests on water offering conclusions as to acidity, alkalinity, ionic condition and dissolved substances may in some instances be important and such conclusions, other than in the most general sense, are best reached in specialist laboratories. Simple field tests for pH values which are readily available in the normal course of soil analysis have direct relevance to the general problems of matching materials and the phenomenon of flocculation of clays which is important in test conditions.

Biological identification and investigation can be important where rampant plant growth or burrowing insects are concerned. Precise evaluation of species by entomologists and biologists is, however, rarely as important as the general procedures for the application of controlling agents – usually toxic chemicals. These are selectively damaging or deadly to various life-forms (including human) and in Europe and North America are generally controlled by stringent regulations on storage, handling, usage and notification. This is not the case universally, however, and substances with lethal or long-term damaging effects are used without contravention of law in many parts of the world.

In Britain the COSHH regulations apply to all substances and practices falling within these general strictures. The procedures and intentions are to be observed in application, and the assistance of manufacturers, laboratories and other specialists can be important. It is a wise precaution to apply similar controls in areas where these may not legally be called for. The impregnation of earth structures against biological attack is technically possible but should be a matter of last resort. Specialist laboratory advice may occasionally be needed.

The commercial application of the results of research has been so important in soil mechanics that the laboratories, their processes and their relevant skills have reached a state of knowledge in soils engineering significantly ahead of the knowledge of conservation. In the civil engineering context, earth structures include massive works of embankment, cutting, tunnelling, land filling, sea filling, erosion consolidation, dam building and land slip. To only a limited extent is the great body of knowledge in soil mechanics applicable to the work of the conservator. There are two exceptions to this – the effects and damage of earthquake and the effects of slump in massive earth structures. In both these areas conserva-

tion advice is likely to demand the specialist skills of engineers in these fields, and unless the conservator has these special facilities at hand, the expetise will need to be introduced. The range of technical analysis and laboratory testing is wide and there is a growing international consistency in the method, evaluation and reporting of results.

The summary below compares the methods of testing and analysis available to the conservator of earth structures.

5.5.i Tests and analyses applicable to porous solid building materials

1. Test: Determination of density. By itself, the density of a sample is of little direct use. Comparative tests can demonstrate the extent of compactibility or compaction in, for instance, block-making by comparing weighed samples with a standard material which is oven-dried. Gaseous methods are now used in laboratories in preference to mercury immersion.

2. Test: Permeability. The ability of a sample to permit the movement of liquids through the fabric without breaking down has a direct bearing on evaporation and ability to discharge water.

3. Test: Porosity. The volume and sizes of internal spaces in a sample are indicators of performance in thaw–freeze conditions and of volumes of water held.

   There can be extreme differences both in pore size and density of pores. While void ratios can be fairly constant in granular material, cohesive soils (i.e. the clay fraction) will produce a great variation. In these, the particle size, and therefore the pore size, is much smaller. Thus, a highly compressed saturated clay may have a void ratio of only 7%, whereas a sodium montmorillonite under low pressure may have a void ratio of 800%.

4. Test: Stability when saturated. Strength and stability in wet conditions are of direct relevance to the performance of a soil and this factor is relatively simply tested in a field station by observing the behaviour of samples in dry, moist and saturated conditions, loaded and unloaded. The test results are comparative and should be related to a base standard, such as the historic material.

5. Test: Dry and wet strengths. Field station tests can give the conservator important information about the strength of soil. Comparisons of known samples and proposed materials can be made by the use of a (pocket) penetrometer, by tests of breakage on samples of specific size and known water content. A wide range of personalized tests can be used to establish the calibre of available material.

6. Test: Atterberg limits. The determination of soil behaviour under each of its four conditions gives important elemental data for use of soils in construction. The Atterberg (and Casagrande) laboratory tests provide the method by which these four conditions (solid, semi-solid, plastic and liquid) may be differentiated reliably, as defined by shrinkage limit, plastic limit and liquid limit.

7. Test: Plasticity and workability. The plasticity of a soil indicates its behavioural characteristics, and these are generally thought of as workability. The tests to define them are essentially practical and may be carried out in the field. They are a very close approximation to the methods used by craftworkers in understanding their materials.

8. Test: Particle size. Particle size analyses are aimed at defining the physical gradation of soil. The sizes of particles as defined by the British Standard 1377 (1975) are as follows: (μm = microns = micro millimetres):
   - colloids (clays and very fine particles of silica, carbonates, etc.) up to 2 μm
   - silts (fine up to 6 μm, medium up to 20 μm, coarse up to 60 μm): up to 60 μm
   - sands (fine up to 200 μm, medium up to 600 μm, coarse up to 2000 μm): up to 2 mm
   - gravels: 2–60 mm

   In the view of some practitioners, the behavioural characteristics of silts justify their classification to include the fine sands up to 200 μm.
   These tests are carried out by:
   - dry sieving
   - wet sieving
   - settlement analyses
   yielding results showing percentage content.

   Susceptibility to frost damage is partially a function of particle size but the behaviour of cohesive soil will be determined as much by geological structure as by particle size.

9. Test: Clay fraction shrinkage. As the shrinkage of clays is the prime factor governing movements in soils in response to changes in water content, the extent of the clay fraction having been determined by particle size analysis, shrinkage tests on the soils are undertaken by volumetric laboratory analysis or linear bar field tests.

10. Tests: Nature and percentage determination of clay types. The cohesiveness and swelling/shrinkage characteristics of clays vary.
    - attapulgites and kaolins expand generally by about 5%, maximum 10%
    - illites generally have an expansion factor between 8% and 11%
    - montmorillonites generally have an expansion between 12 and 18%, but it can rise to 24%

    The tests are either laboratory analyses determining chemical content (silicon, aluminium, etc.) or simple field tests estimating proportions of clays in smectite mixes (Emerson test).

11. Test: Nature and percentage determination of soluble salts. Salinity is a measure of impurity in soils; some salts are more troublesome than others. By practical definition they are water-soluble. Their damaging effects are largely related to deposition within pores, but they may also be inhibitors or even reactants with materials introduced for other purposes.

    Field testing is relatively useless in giving real precision to the salts present.

    Laboratory analyses are only of value if the salts are defined by type and weight.

    Results giving percentages of elements (e.g. Na, Mg, etc.) are of little value to the conservator.

12. Test: Nature of organic matter present. The objective of tests of organic material is to demonstrate whether decomposition has produced or will produce materials of acidic or chelating nature which will affect the distribution of metallic ions within the soil, and particularly within the clay fraction. Interpretation is for the specialist

and the analytical work is sophisticated in all but the simplest examples.

13. Test: Dating by inorganic methods. Sophisticated analytical methods are now in use and others in development which allow the determination of age of material and date of burial. These depend on measuring rates of decay or of accumulation of effects of radiation. They are all expensive, delicate and approximate in result. In consequence, they are likely to be confined to museology rather than conservation. Some organic methods such as carbon dating (also a decay method) and dendrochronology may have occasional relevance to historical enquiry.

14. Tests: pH value and carbonation. High acidity and high alkalinity in soils are to be avoided and are indications of instability or impending change. Tests of litmus (now disfavoured) phenolphthalein and other colour change indicators are available and suitable for use in the field station. A neutral (order of pH 7) condition is to be aimed at. Secondary effects can be valuable – the use of phenolphthalein to determine the extent of carbonation in lime mortars or lime-stabilized earths is produced by alkalinity. Calcium hydroxide displays itself by a highly alkaline reaction. The carbonated lime, being neutral, shows up much less, and hence the degree of conversion or 'set' can be determined.

15. Tests: Chemical analysis. The utility of chemical analyses varies enormously, partly in relation to what the conservator seeks. The chemical and geological 'fingerprint' (contents analyses) can demonstrate origins of materials which might be crucial historically. Where the analysis can focus on molecular structure rather than on elemental breakdown the results are nearer the needs of the worker in the field. It is important therefore to have sufficient background to be able to determine what results are required. To know the availability of free metallic ions may be far more helpful than determination of the presence of rare elements, subject to sufficient knowledge of, for instance, the nature of change in clays due to ion exchange.

Chemical analysis, other than of the simplest kind, is a laboratory matter and therefore involves other skills.

16. Test: Geological classifications. Soil particles are only identified approximately in the field, with some obvious important exceptions – silica, chalk, mica among them. Microscopic examination greatly extends the range of knowledge. The geological classifications of material are of the greatest assistance in determining the nature and behaviour of the soil mixture. The weight of knowledge of material types and behaviour is such as to make this aspect a prime resource for the conservator, reinforced by the Emerson test (see above).

17. Tests: Biological analysis. Plant life primarily comes under scrutiny to reveal active or potentially active organisms which may have an effect on the structure, or inactive and decaying material which may produce unwelcome effects. Generally, the common understanding of biological behaviour is sufficient but in exceptional circumstances it may be necessary to know whether a particular organism will cause an effect to be welcomed or safeguarded against. Historically very sophisticated analyses, such as pollen identification, can provide important evidence but at no little expense.

18. Tests: Scanning electron microscopy, diffraction and spectroscopy. The scanning electron microscope can trace shapes at even molecular level, allowing physicists to interpret or predict phenomena extremely delicately.

Crystal structures and the elements in a material are determinable by diffraction of X-rays and other electromagnetic wavelengths.

Fascinating as the results may be, there is rarely any occasion for the conservator to call for such material, but where it is available for other reasons the information should be combined with other available results for possible application to materials studies.

5.6 Purposes of testing and analysis

The objectives of testing and analysis should be clearly identified in the conservator's mind. While field-testing may be absorbed within the general costs of conservation work, laboratory analyses and laboratory testing are usually separately identified and should be undertaken purposefully. Only those tests whose results,

Figure 5.3 Electron microscopy.

While of no day-to-day importance electron microscopy can reveal, in research situations, important evidence of the formation and relationship of particles, most usefully in magnifications in the ×500 to ×2000 range. In this instance the larger crystals are gypsum and the smaller particles are air-deposited carbon. The electron microscope can achieve magnifications in excess of 90 000 000 actual size.

either positive or negative, will contribute to decision-making and materials selection should be commissioned. It is as much a failure on the part of the conservator to make no use of the results of the care, skill and investment of effort of laboratory colleagues as it is to make the wrong decisions through having failed to commission such testing.

The purposes of testing can be summarized under three broad headings:

- an understanding of the soil structure in the existing monument – divisible into present and original condition
- evaluation of materials for proposed intervention – methods of application and anticipated results
- understanding the long-term effects of an intervention – monitoring and final analysis

At the outset of a conservation programme the conservator should design and record the intentions of the analyses and testing which will go with the programme. The discipline of making such design decisions is in itself a check upon their purpose, quality and quantity and moreover assists in budgetary control. It may have the secondary benefit of achieving the necessary budget by justifying the programme

and of presaging the publication of the results. The objective should be that the programme is on the one hand comprehensive and sufficient and on the other neither wasteful nor inadequate. There is a wide range of problems and soils in the conservation of earth structures and an extensive range of available materials, some of which are little tried and others emergent on the market. Careful evaluation and research into experience gained elsewhere can lead to the use of techniques which are more advantageous, economical or effective in the long term than readily available alternatives. The serious conservation of earth structures by means other than purely traditional techniques is a very young art. Expertise and experience are concentrated in localized groups and individuals with special skills. As in any emergent area of conservation activity, initial conclusions and deductions are often tentative and many are proved inadequate or wrong. New materials are tried and the complexities of varying problems are progressively understood. A programme of testing and analytical research therefore must be coupled with a careful programme of enquiry into parallel work elsewhere if the best results are to be achieved and advantage is to be taken of the most recently gained knowledge and experience.

6

Inorganic materials in conservation and repair

6.1 Traditional background

Conservation aims to make good damage as it occurs; repair provides interventions to replace substantial loss. Stabilization strengthens material *in situ*: the various forms of organic stabilization are described in Chapter 7.

It is an essential objective of repair that a structure shall be left sound and stable on a long-term basis with a minimum of intervention, having been handled in a manner sympathetic to its intrinsic qualities. No better technique can be recommended than is provided by traditional practice, even though the traditional method may have technical inadequacies and be labour-intensive. Continuity of tradition is the very stuff of history and any action of conservation will carefully look first at all the ramifications of traditional methods, continuity of skills, social stability, local employment and local pride. Methods vary widely. Their continuity and their recording are integral to the work of conservation, even though modifications in technique may be appropriate. These must be introduced with care and inappropriate techniques may require firm rejection. Some earth structures may, for instance, carry in-built scaffold – supports projecting stones or wooden bars. To provide steel for this purpose might be both easier and cheaper but the arguments against doing so are unimpeachable.

In general, the visual qualities of the surface of earth structures are relatively fragile in the sense that they are not readily or mechanically simulated. They are fragile too in a structural sense. The work of the conservator, therefore, must be to provide, in addition to structural stability, a surface comparable to that provided by traditional finishes. These have usually been additive – that is to say, each successive finishing layer has been built up as additional application until the surface quality becomes dependent upon the visual effects of the additive process itself. By stripping off earlier coverings, paint, tar, limewash or mud render, an essential part of the quality is lost, to be recovered only by sensitive replacement. Where traditional methods of finishing are employed, such stripping down is unlikely to be necessary except in cases of drastic repair, but where artificial consolidants are to be used the case is very different.

6.2 Inorganic interventions

6.2.i Juss and limes

Inorganic consolidants of earths have been known in practical terms for a very long time and in their most basic form – lime and juss have been used from time immemorial. Calcite is a natural component of earths, but lime and juss, being artificial preparations, must be treated as synthetic consolidants added at the stage of preparation of the earths. Juss or fired gypsum generally has been used only in dry climates as it is slightly water-soluble. Earths consolidated with this material are usually prepared from gypsum-bearing earths but sometimes fired material (plaster of Paris) has been added at the stage of preparation and mixing prior to forming block or brick. The techniques are often rugged or rudimentary

Figure 6.1 Bid-Bid, Oman. Repair using traditional techniques ensures a continuity of the plasticity and softness of contour that is the essential characteristic of earth construction.

and once placed, the material sets by some recrystallization. By comparison with the untreated earth the final colour is appreciably whitened by even relatively small amounts (up to 2.5%). With additions as little as 1% the finished material takes on a set. Above 5% it effectively behaves as a weak plaster and above 10% concentration it may be regarded as being capable of taking on a plaster finish which, in ranges of upwards of one-third of

the addition of juss, is capable of being brought to a polish. Gypsum is entirely compatible with lime as a plaster-type material; the effects of hydrated lime are to provide a softer and more workable product without the sharp setting time which is normal with higher concentrations of juss. The additions of lime putty, both hydraulic and non-hydraulic, have the same effect of diminishing the short setting time and they ultimately produce a harder material. Even the hardest such render or block, however, will always be placed in the category of a soft plaster which can be cut with a knife.

The use of Portland cement as a further additive in a soil modified with juss should not be considered. In a dry situation it can offer no advantage and in a situation of sustained or intermittent wetness, gypsum-based additives are an addition unlikely to be considered by the conservator due to their own solubility. In combination with Portland cement, complex long-term chemical reactions are set up with the production of soluble calcium chloride. This is disastrous.

There is no published understanding of the performances of many different types of soil in the presence of gypsum sulphate and its derivatives, particularly in respect of the different types of clays, whose crystalline behaviour is likely to be modified by even small amounts of calcium sulphate capable of entering solution. The silt and sand particles are less likely to be affected chemically and physically by the behaviour of gypsum in recrystallization and in the present state of knowledge a prudent general rule will be that a soil washed clean of clays and organic materials but retaining silts and sands will be a more reliable base for gypsum additives than a soil straight from the ground. Where reliable local knowledge and long-term experience are available, their application, coupled with careful testing may be a sufficient basis for a conservation judgement. As with the addition of limes, the availability of calcium cations originating from the gypsum cause saturation of the clays (in the electrochemical sense). They enter and are retained within the platelets of the clays and may change their composition. They may also produce calcium alumino-silicates and similar crystal-forming compounds which can appear quite rapidly and give a cementing effect to the earth. These reactions occur in a highly basic environment, upwards of pH 10.

The use of lime as a stabilizer of earth structures is an ancient tradition and is, therefore, admissible as a non-synthetic – although artificial – technique. Limes vary. Hydrated lime carries a smaller proportion of available calcium hydroxide than freshly slaked lime putty, having already absorbed some carbon dioxide from the air. Hydraulic limes carry

other minerals – magnesium, aluminium and silicon oxides – which also produce combinations with crystalline potential other than the simple formation of calcium carbonate, which is the basic mechanism of crystallization and therefore of the solidification associated with pure lime putty. Limes used in mortar mixed with clean water and sand harden by the absorption of carbon dioxide from the air by the prime mechanism of the formation of crystals of calcium carbonate. This is a slow process. In soils, carbonation of the lime remains a prime mechanism but where clays are present the reactions are believed to be dependent upon the activity of clay particles in which calcium ions find lodgement in the vastly greater surface areas offered by plate structures. Therefore, the nature of the clays has a significant effect. The cations in a solution containing calcium (and also magnesium) oxide saturate the clay minerals. The clays have properties which derive from the charges on their surface areas and these are modified in reaction. It has been suggested that the production of aluminium silicates and aluminates by calcium hydroxide creates cementing bonds in significant concentration. These, allied with calcium carbonate, affect the stabilization of particles and produce an initial rapid increase in strength which then improves with the production of calcium carbonate crystals over a longer period of time. The stabilization phenomena have been divided into fast and slow reactions. In the former, the reaction involves exchanges of cations – absorption of calcium hydroxide molecules and concentration of ions – whereas the latter involves silicous, aluminous and ferrous cementations succeeded by the locking in of calcium by carbonation. These processes are complex and depend on the nature of soils, the water, salts and temperature. Alkalinity is generally accepted as an indicator of the degree of carbonation: a fully converted lime is neutral, in contrast to the actively basic calcium hydroxide which can be detected by the use of phenolphthalein, which provides a reversible coloration in alkaline conditions. The test can be used in soils but due to the coloration, a white background is needed. Other soil constituents may in any case mask or distort the test, so it is of only moderate value.

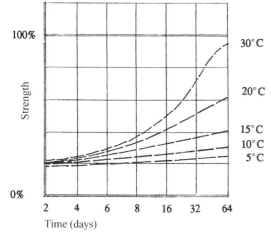

Indicative graph for hardening of soils with 5–15% mixed clay content stabilized with 3–5% non-hydraulic lime under adequately moist conditions on varying ambient temperature conditions.

While the general processes of change occur progressively, the occasional formation of alumino-silicates and similar materials can occur rapidly. No attempt has been made to include these variable phenomena in the graphical representations.

Figure 6.2 Rate of hardening of lime-stabilized soils.

6.2.ii Cements

One further common stabilizer must be considered. Portland cement is a most effective building material which has been in increasingly wide use for a century and a half. Because it is so effective its manufacture has become universal and it is freely available throughout the world. It has been widely and frequently used in the stabilization of soils. Like all materials which are widely available, relatively cheap and effective in application, it has been misused, and misuse has brought with it an entirely justified fringe reputation of inadequacy and unsuitability in many historic building applications. Other cements, not strictly Portland cement, are sufficiently similar to be included within these generalizations.

The setting of cements is modified by organic impurities, organic acids, oils, sugars and esters (organic salts) as well as many of the salts commonly occurring as pollutants in earths. Other organic materials such as resins may be compatible with cements, and may enhance their performance.

In cement mortars and concretes the reactions which produce satisfactory finished materials are complex but relatively consistent. They are based on the formation of long needle-like crystals which interlock to form a matrix around the sands and aggregates comprising the bulk of the mix. In earths the reaction is even more complex. There are even more potential reagents and the proportion of admixture is far smaller in terms of the cement component. In stabilizing earths the objective is not merely to substitute earth for sand and gravel in a concrete mix: it is to maintain the earth as the overwhelming constituent with a sufficient admixture of cement to provide rigidity and resistance to degradation under high moisture conditions. These objectives can be achieved at the expense of brittleness and a significant colour change towards greys. The resultant material has many disadvantages other than its unattractive colour and lack of flexibility. Soluble salts can be produced and these, drawn to the outer surface, can be disfiguring and damaging as they interfere with the structure of the clays within the soil. Crystal formation also interferes with clay structure and can negate the cohesion provided by the clay component in favour of a locking of particles at their surfaces. Used in an earth mix in substantial volume, cements produce a material too hard to be compatible with native earth structures.

These disadvantages are sufficient to put great limitations on the use of cements with earths. However, advantages can include a considerable reduction in shrinkage during setting and a relatively fast setting time. The reduction in shrinkage derives from the formation of crystals within the interstices between clay particles. The successful use of Portland cement as a stabilizer in earths has generally been in one of the following limited areas:

1. Foundations where the native earth may be subject to saturation and plasticity. In this circumstance the admixture of 1–3% by dry weight of Portland cement can produce a more reliable foundation for an earth wall. In moist soil conditions an inhibitor to prevent rising moisture carrying salts up the wall from the stabilized earth is an essential part of the construction. While

damp-proof coursing in the conventional sense is not helpful – providing too precise a demarcation – a form of construction which inhibits rising moisture is essential. Provided that sulphates are not present in quantity a foundation of relatively impervious stone bedded in a cement mortar can have the right effect. Alternatively, dry-stone walling is suitable in dry and moderate climatic conditions.

2. Capping of archaeological remains and similar structures where the cement-stabilized earth mortar is not subject to fluctuating live loads. To succeed with this technique, a carefully considered design will include thermal movement joints at wall junctions and will divide long runs into short lengths. In this instance a strong mix incorporating between 5 and 10% by dry weight of Portland cement may offer the most advantage. Cappings should be of adequate thickness and should not incorporate ferrous reinforcement. Drips should be carefully detailed; it should be borne in mind that water emanating from these cappings will carry metallic (particularly calcium) cations producing salts depositable in the wall. Soil/cement cappings of this type will not bond fully to earthen substrates. They can be designed to be stable as a dead load or, if this is not feasible, a mechanical bond may be required. The use of joggles in joints and simple tongues and grooves is recommended as a first recourse.

3. Load-bearing components. Earth blocks stabilized with cement and manufactured to different specifications of compression as well as composition are widely used in the drier tropical regions. Their structural and other physical properties have been extensively tested and in some areas – Mexico and New Mexico, for instance – they are well-established commercially. Elsewhere their manufacture and performance are less vigorous and they are available to order rather than as a stock item. The conservator may use cement-stabilized block as a compatible structural component where substantial infilling is required for its advantages of stability, minimal shrinkage and predetermined performance characteristics. Earth–lime mortars and cement-stabilized earth mortars are compatible and should be designed to be weaker than the blocks.

4. There are circumstances in which the conservator may choose to construct stabilized earth lintels over openings in circumstances where steel, concrete and timber would be inappropriate. A structure shattered during an earthquake or by some other calamity might be reconstructed using cast-*in-situ* reinforced stabilized earth, particularly where a further facing of render is to be applied. Such a construction would be less likely to exhibit the weaknesses of cracking and excessive point loading which would attend the use of more rigid materials.

6.3 Modified earths used in repair

Since gypsums, limes and cements come to the building site as an insoluble powder they cannot be carried into the material by a solvent. Therefore they must be mixed with pulverized raw earths and then wetted to achieve setting. This inhibits their use and predetermines their application via newly constituted earth material. It is a fundamental limitation.

Several methods of use are common to gypsum, limes and cements, each being appropriate to differing circumstances according to the base materials and local circumstances. Their uses relate to the preparation of grouts, mortars, fillers and substrates in the repair of indented or damaged surfaces.

To varying degrees the shrinkage factor in these modified earths is improved and, given the advantage of identifiability in their use, there can be circumstances in which their use is to be recommended for injection into voids, cracks and the interstices of failing structures. Sometimes an entirely fluid injection is required, i.e. one which will penetrate under hydraulic pressure (usually a head of less than a metre with application via a small tank and plastic pipe). In such circumstances a synthetic grout employing fly-ash, fine sand and possibly small amounts of Portland cement may be considered. The modified soil mixture is more likely to be used at the consistency of a mortar as an inert filler.

However, where it is important that there is no shrinkage effect or there is a requirement for a significant bond between the filler and the base material, such amended mortars are unlikely to be thoroughly satisfactory. The particular use of such modifications in the formation in substrates depends on their being able to provide a bond and adhesion to a further coating. Much earth construction is faced with historic plasters and lime- or cement-based washes or renderings. Generally these have little or no adhesion to an earth structure other than a moderate mechanical bond. Differential movement can reduce or eliminate this bond. In circumstances where it is important to tie the surface finish to the wall more reliably than has been the case previously, a modified earth may prove the answer. Blockwork or backings formed with gypsum, lime or cement additives may be put in as a substrate for a finish, with the expectation that a reliable bond will be formed.

6.4 Modified muds used as filler

A grout-filling technique in earth structures depends on a specification of a material which will not shrink on loss of water. It also depends on techniques which do not induce shrinkage in adjacent mud construction. If the clay component of the earth is of an expansive type, any technique which wets the clays will induce a shrinkage crack as they dry again. Any technique which does not involve wetting the clays will cause the grout to lose water on contact with the dry earth, upon which the grout will coagulate. Any technique involving such wetting as will overcome suction risks liquid collapse in the historic material.

Despite these practical difficulties the introduction of modified mud grouts offers advantages which justify their use in certain cases. In earth structures, even under pressure, no grout can be relied upon to penetrate cracks less than 38 mm in consistent width due to coagulation. Small voids will not be filled. However, the advantages are that the structure is undisturbed and penetration is not dependent upon excavation of the cavity. A secure, solid and coherent fill may be obtained in a deep irregular void which occurs in conditions of settlement and earthquake damage without further structural interference, and the irregular shape of the cavity causes a mechanical bond which can restore strength.

Experience shows that muds modified with lime and pulverized fuel ash (PFA) produce the right consistencies and avoid the shrinkage problems associated with unmodified material since the larger particles provide a resistance to shrinkage. There may be occasions when additional reinforcement is required for linkage, in which case resin/fibreglass rods may be set in cavities formed in the sides of the failed zone; the pockets around them are filled by the grout. Alternatively fired clay pipes and tiles have been used for the same purpose.

A formulation of grout will depend on the soil of the structure. Modifiers may include Portland cement, hydrated lime, lime putty, PFA, brick dust and gypsum (to provide an expanding set). Sands may be required. By experimental determination a mixture should be obtained which will set without shrinkage, will flow readily through a 2 or 3 mm void, will be of density, hardness and strength similar to the structural earth, and will not settle out when standing prior to setting. The important characteristic of PFA is that its particles are globular, having been formed from melt in air-suspension. Consequently they move more readily through small voids. They also have a pozzolanic or cementitious effect with lime (see below). A set may be designed to take less than a day or several days.

Proportions may be in the order of two-thirds soil, one-sixth sand or finely powdered chalk and one-sixth lime with a pozzolana (cement, PFA, brick dust). Experiments may then produce variations in which the proportion of soil is reduced while the other materials are a constant. The soil type may also be changed. Only by careful evaluation of results in relation to local circumstances will the most appropriate results be obtained. Gypsum will be used only where water penetration is a minimal problem and where an expansive filling is required. Careful judgement will be needed to determine comparative mixes and times of application in relation to setting. Where such a bond is required the setting of the substrate should be approximately half complete at the time of application of the surface or further overlayer. The ideal theoretical situation would be a continuously varying concentration of the modification material so that no boundary layer is formed. In practice this is not achievable but the conservator may well decide to use a series of successively richer applications or coats to simulate such a condition. There is evidence that the creation of highly polished gypsum plaster surfaces on earth cores was achieved successfully in part using the multiple application method with increasing richness as the surface is approached. Such operations, however, often included other materials such as albuminous animal products and comparisons should, therefore, be made with caution.

The construction and repair of floors in earths are particularly relevant since they have, through the ages, been constructed with hardening materials and animal products to the point when it may be impossible to distinguish between the original construction and materials incorporated during use. Few native earths produce a floor which will stand up to substantial wear and in historic buildings floors are frequently given much higher levels of wear than their builders would have anticipated. The introduction of stabilized earths, therefore, may well be an attractive answer in terms of conservation.

The conservator will weigh integrity, nature of original materials and the type of finish required against circumstances such as soil moisture content and levels of wear anticipated. The extent to which soils are expected to breathe in flooring and the requirement of animal-based consolidants will be assessed in the light of available materials, including the organic consolidants discussed below, the objective being the material of greater cohesion than a natural earth but with similar characteristics of flexibility and appearance. Analysis will provide understanding of the nature of the original material from among a wide variety of possibilities. One very significant example is a floor conserved in northern China at Dadiwan in Gansu province. The floors in the archaeological sites date from the Yang Shao period, a neolithic culture approximately 5000 years old. The floors were manufactured from a form of fired loess impregnated with calcium carbonate. The combination of montmorillonite clays, sands

and calcium carbonate is not dissimilar to the basic materials employed in the manufacture of cements and the firing had been carried to a sufficiently high temperature to calcine the deposit. The floor is, therefore, a primitive concrete, perhaps the earliest known.

While this is an extreme example illustrating the breadth of range across which these materials are distributed, the conclusion must be that analysis may yield evidence of cementitious compounds at use in all periods of history and the use of equivalent consolidants in replacement material may embrace extremes in the range.

6.5 Manufactured materials as additives

Totally synthetic (i.e. of manufactured products as opposed to natural) materials are available as an alternative to consolidated or modified earths. Of these, the principal is fly-ash, PFA the byproduct of burning pulverized coal in air suspension – fluid bed combustion. The inorganic residues (ash) in the coals are melted in intense heat, solidifying in the flue gases as tiny, spherical globules of glassy slag. They are available in large quantity and contain few pollutants other than sulphur which occurs in varying degrees. Only those supplies which are virtually sulphur-free are suitable for use as fillers. The surface of these very small granules, consisting very largely of a cooled melt of slag – an alumino-silicate – will have a pozzolanic reaction with calcium hydroxide, producing a hydraulic set with limes with which it is mixed. The same is true of other sintered materials such as brick containing active silicate. An additional merit of fly-ash is the ease with which its particles will move through porous spaces due to their shape. This material, of relatively recent introduction, has proved extremely useful as an inert filler and low-cost constituent of grouts being used in voids in buildings and on a massive scale in forming embankments, road bases and in other forms of civil engineering work. It is of varying availability and diverse sulphur content.

It can also be used neat as a simple filling material in earth where it has the merit of setting off in an aqueous mixture to produce a rigid structure of modest cohesion but substantial compressive strength comparable with many earths. Because its moisture absorption is low and the areas of contact between particles tend to be confined to single points, its movement on take-up and loss of water are minimal. Particle sizes run through the sand and silt ranges and the percentage of clay size particles is very small – there are virtually no grains below 0.1 mm in diameter. Fly-ash in consequence can be placed as a grout with important properties of penetration and with avoidance of shrinkage on drying. The inclusion of small amounts of cements produces considerable increase in set. This latter quality is significantly increased by the presence of sand. In the presence of pure lime fly-ash gains strength by the formation of crystals incorporating silicates in addition to the carbonate crystals.

Useful as it has proved, conservators may have to bear in mind that PFA is a material of the moment. Its availability and cheapness are products of current methods of power generation which may diminish quite rapidly. Proprietary mixtures containing fly-ash, cement, sand and some other materials in careful proportion are available on the market and are a reliable source in the quantities required for conservation. In many instances, however, their design strength will be for their intended use as a grout in brickwork and masonry and this may be significantly stronger than is compatible with earth structures. In attempting to weaken such materials particle size must be considered. Since the range of particle size in fly-ash tends to be in the order of 1 mm, coarse sands introduced in the mixture can separate out within the pores, acting as a filter, and allowing the fly-ash to continue in concentrated form into further interstices. Fine-ground hydrated lime and fine sands should, therefore, be specified to minimize segregation. Where strength between an earth structure and a fly-ash-based fill becomes important proprietary products may be less satisfactory than direct supplies of the material modified with small amounts of lime, cement or sand as judged necessary. A significant characteristic of the material is its grey colour – a hard metallic grey comparable to a dark Portland cement. While this can be very useful in tracing progress of the material through walls it is disadvantageous in general

aesthetic terms. The colour range can be partially modified through browns, reds and ochres using mortar pigments or powdered brick which is entirely compatible with PFA.

Typical material contents are by weight:

Silica	50%	SiO_2
Alumina	25%	Al_2O_3
Iron oxides	10%	FeO_2 etc. occasionally much higher
Calcium oxide	5%	CaO
Magnesium oxide	2%	MgO, with smaller proportions of sulphates and alkalis

Turkish conservators have experimented with the use of mixtures of PFA, brick dust and lime, with results that yield encouraging figures in terms of shrinkage, stability, strength and thermal transmission, while illustrating the control over colour, the relative economy of the material and its practicality in terms of gap-filling and block-forming. The range of properties achievable suggests that it can be modified to make it applicable to the repair of earth structures and of brickwork. It is an interesting reflection that, although PFA in itself is a new material, the tradition of using mixtures of lime, powdered brick and ash is long established both as fillers and in the provision of durable raw materials.

The availability of brick dust is itself a reflection of technical methods. Highly sophisticated modern works produce little or no useful brick dust. The more basic hand and mechanical techniques used in countries where labour is relatively cheap give rise to substantial volumes of this very useful byproduct.

Subject to adequate information on the extent of sulphur, the design of inert fillings using brick dust, lime and PFA in various admixtures with other possible additives has the merit of being flexible, of using cheap materials and being independent of sophisticated processes of application.

6.6 Consolidation by other inorganic materials

Although the principal reagent in the formation of calcium carbonate crystals arrives in gaseous form, i.e. carbon dioxide and water vapour in the air, requiring only the presence of the free calcium cation in solution, the only artificial use made of the introduction of gaseous carbon dioxide is in laboratory conditions. Experiments in accelerated carbonation in earth structures have been carried out, notably in Brazil at the University of Bahia. In these experiments the objective was to study the carbonization process and the secondary production of aluminium silicates and similar compounds. Elsewhere experiment has been carried out with water enriched with carbon dioxide. These experiments relate to stone repair and have not yet been applied to earths.

Much of the chemistry of crystallization in soils revolves around silicates which are capable of producing complex ranges of compounds with multiple cations of the reactive metals such as potassium, calcium, aluminium and occasionally magnesium. Success has been recorded in the use of potassium silicate which is highly soluble in water and is, therefore, carried into earths in solution. This material has been known in the past as a sealant of, for instance, eggshell and has therefore been used to preserve eggs under the name isinglass. It is used as a medium carrying pigments for retinting brick surfaces where there has been a colour-match problem and as the basis of durable inorganic paints. For consolidation, potassium silicate is preferred to the cheaper but similar sodium silicate because it does not produce a surface scum or efflorescence. Although success in its use as an earth consolidant has been described, it is clear that the process depends on careful and knowledgeable execution backed by skilled analysis. The results have yet to be proven by extensive field trials and sustained weathering. The methods of application are straightforward and the materials are simple and readily available. Furthermore they produce in the consolidated earths an inorganic structure comparable to the formation of the soils themselves without the inherent problem of complex organic compounds liable to biological attack and breakdown.

As the process of consolidation concerns the internal structure of the clay particles, preparation of the material must be judged in relation to the types of clay involved. Crucial studies have concentrated on clays containing a high proportion of montmorillonite – the most expansive of typical clay materials. Research has been carried out at the Dun Huang Academy in Gansu, China with the assistance of the Getty Foundation. After unsuccessful experiments with polyvinyl acetate and solutions of sodium silicate, success has been achieved with potassium silicate solutions in which the critical molecular (molar) ratio is established as between silicon dioxide and potassium oxide. The ratio is critical because a surplus of potassium oxide remaining will absorb carbon dioxide from the air and form potassium carbonate, leaving unattached highly alkaline silicates. If these approach acidity, colloidal silica is precipitated. Consequently pH values are necessarily high in this operation, and strict control of the chemistry is important.

Laboratory trials establish the appropriate molar ratios by experiment, the samples having been tested for stability, resistance to water and carbon dioxide. Calcium fluorosilicate has been used as a solidifying agent and aluminium silicate as a cross-link. Application by multiple spraying carried out in basic conditions (pH = 10), reinforced ultimately by analysis under scanning electron microscope, has demonstrated the formation of an interlocking crystalline network providing rigidity within the clay plates and consequent inherent stability in the treated material. A full range of analytical tests has vindicated the results and the consolidated material has not discoloured. These promising results have only been obtained under carefully calculated conditions of material preparation and circumstances of application. For some time to come the process is likely to be confined to significant archaeological material where the work is supported by a fully equipped laboratory service, as the results are dependent upon careful control and the determination of precise molar ratios in the solutions.

Under most circumstances sodium and potassium silicates available commercially are forms of glasses and are thus amorphous. Crystalline forms probably have an excess of K_2O.

6.7 Semi-organic consolidants

A longer-tried method of introducing similar consolidants involves the subtle technique of depositing silicate crystals by the use of organic salts which decay by polymerization to leave the silicates in position. The first stage in the reaction is evidently the precipitation of a silica gel which may then react with other molecules to form silicates. The silica gel is a 'glue' in its own right and can be used to bond a wide range of aggregates. Proprietary formulations are available for consolidation of stone, in which a ketone carrier contains tetra-ethoxysilane and methyltriethoxysilane plus a catalyst. The active constituent, ethyl silicate, reacts with water in the presence of the catalyst (dibutyltin dilaurate). The chemical process goes through several stages and can follow several paths, different products being produced in partial or total hydrolysis. The reactions can be spread over a long period and there are reported to be advantages in the slower processes of polymerization. With the discharge of ethyl alcohol and water a silica gel will be formed. Within an earth matrix this gives way to regular arrangements of silica within the clay particles. With the ultimate complete polymerization the structure and porosity of the earth are restored while the clay particles are internally stabilized. This stabilization is sufficient to give rigidity in conditions of normal weathering and greatly to reduce the speed with which weathering takes place. The treatment is not reversible, however, which is a minor disadvantage countered by the advantage that it does not change the colour or characteristics of the earth. Methods of application vary from irrigation through bore holes to spraying, and in consequence depths of penetration and distribution concentrations vary. Optima have not yet been fully established as receptivity varies from circumstance to circumstance. Of all synthetic materials introduced as an earth stabilizer, ethyl and methyl silicates are as the most successful to date.

An advantage of the treatment is that applications of other materials to improve water resistance remain possible. Since ethyl silicate is simply a consolidant there may be circumstances, particularly where important finishes are involved, in which additional strength is required. Other gap-filling and gluing materials can then be applied on the completion of the polymerization of the ethyl silicate. In areas of high salinity materials can be consolidated prior

Professor A. V. Hofmann in 1861 proposed to a meeting of architects in London the use of 'silicic ether' for the conservation of stone. The material remained rare for half a century until its use for stone consolidation was patented in 1926. Its use in earths has taken place since World War II.

$$\begin{matrix} & OH & & & OC_2H_5 & \\ & | & & & | & \\ OH- & Si & -OH & \quad H_5C_2O- & Si & -OC_2H_5 \\ & | & & & | & \\ & OH & & & OC_2H_5 & \end{matrix}$$

othosilicic acid which becomes, by substitution, tetra ethyl silicate.

In long chain form ethyl silicate polymers will terminate with one OC_2H_5 radical being substituted for the oxygen bond on the soil particle as below.

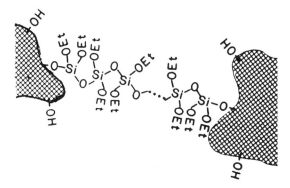

Et = the Ethyl radical ($-C_2OH_5$ or $-OCH_2\ CH_3$)
Si = silicon, OH = the hydroxyl radical

Binding is achieved initially by a polymer chain whose effectiveness is dependent upon the availability of the OH radical bonded to the surface of the soil particles. In earths these are abundant. Silicon links permanently to the oxygen atom in place of one of its hydrogen atoms. The ultimate bond is silicon–oxygen–silicon etc.

In subsequent reactions the ethyl radical modifies and disappears by evaporation, leaving the silica chain in position. (The methyl radical may be substituted for ethyl in general principle, providing a parallel system of consolidants.)

Figure 6.3 Ethyl silicate bonding.

to treatment or additional strengthening, and even prior to reduction of salt content – a process considerably facilitated by the additional strength of the consolidated material.

Soils may be composed of mixtures in any proportion.
To obtain absolute values, apply a reduction factor proportionate to the sums of 'read-off' values

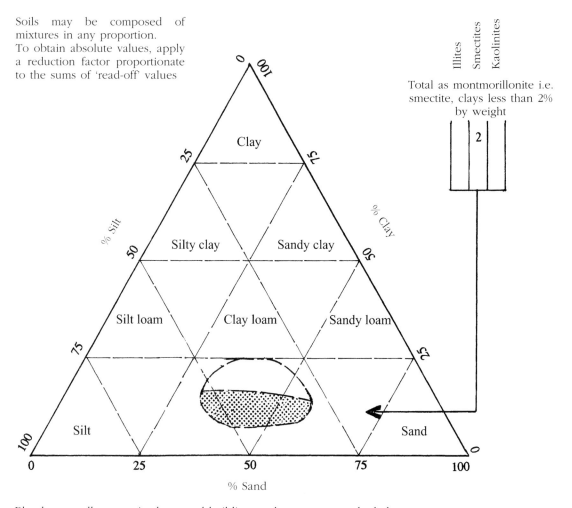

Blends generally recognized as good building earths: target area shaded.

Recommended soil mixes for earth block and mass walling will have less than 5% organic matter, less than 2% solubles, not more than 15% clay, should achieve not more than 18% water absorption, a density approaching or exceeding $2000 \, \text{kg/m}^3$ and a dry strength of $2.25 \, \text{N/m}^2$.

Figure 6.4　Soil composition and structural earth mixes.

The use of ethyl silicates has the disadvantage of greater expense than the totally inorganic materials and, like them, requires careful preparation and evaluation of the applied materials and methods. Careful testing of samples makes the process lengthy and in consequence the applicability is likely to be confined to significant archaeological and important architectural elements.

6.8 Guidance notes for mixes of natural materials used in repair of earths

Salvaging and reconstitution of previously used unadulterated material are generally desirable. The salvaged mix should be analysed and matched unless it is evidently polluted, impure, otherwise unsuitable or

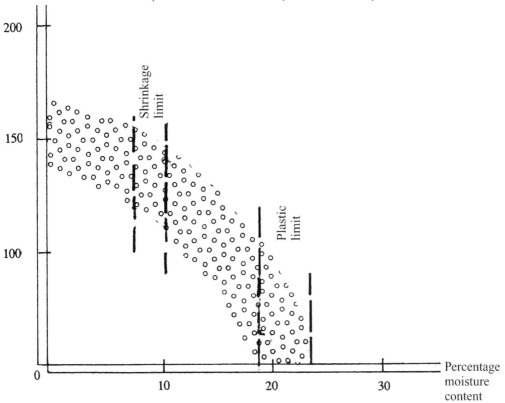

Compressive
strength
kN/m²

Traditional earth/fibre mixes related to varying moisture conditions. The parameters present the zone in which lie typically satisfactory mixes, with 10–20% mixed-clay content and 1–1.5% by weight, fibre content (normally straw). Ideal mixes lie close to the centre of the band. *Note:* many earth mixes lie outside these parameters but are widely and successfully used.

200

Shrinkage limit

150

Plastic limit

100

0

10   20   30

Percentage moisture content

Figure 6.5 Strength/moisture relationships in fibre-reinforced earths.

disadvantageous. Proportions are by volume. Dung is from ruminants. Where identification is required for historical clarity a modern material may be introduced.

6.8.i Mud renders (arid climates)

Mixes should be a range including clays not exceeding 25%, silts up to 45% and fine sands up to 20%, coarser sands up to 10% with up to 5% by dry compressed fibre as finely chopped straw and/or up to 5% dung, all well-mixed until plastic, having matured for not less than 24 hours before being reworked and applied.

Clays should be evaluated to ensure that they contain only a moderate proportion of

highly expansive material (smectite ranges). They should be tested to discover the extent of contraction on drying. If the material cannot provide moderate characteristics the clay proportion can be reduced in favour of silts and sands. Alternatively, less expansive materials such as PFA may be introduced or the clay may be modified by the introduction of hydrated lime, best added during the initial mixing. Colour changes will result from these additions. The proportions of fibre may be increased to compensate for the reduction in clay content. In traditional mixes the fibre is sometimes introduced in the masticated form of dung, perhaps supplied dry. Renders used in wetter climates may contain increased proportions of lime to give greater resistance to damp (i.e. set) and improved bonding to a limewash or other coating.

Apply as a soft render smoothed on to a surface sufficiently well-dampened to reduce suction. The substrate may be keyed physically.

6.8.ii Daubs

Include hair-type fibre, or equal proportions of chopped vegetable fibre down to average lengths of 30–50 mm added to at least six times their volume of sand:silt:clay mixes, evenly graded and containing 10–25% clays. The fibre may be greatly reduced in proportion, depending on type. The mix is brought to a thin doughy consistency, sticky enough to cohere in lumps and hang on the tool. It is delivered with force in small amounts against a supported panel or by equalizing deliveries on opposing sides of the panel.

The mix should be allowed 24 hours to mature. Hydrated lime or powdered chalk may be included up to 5–10% by volume. A modest addition of dung from ruminants may be included to improve workability. If a bonding agent is applied to the sticks or wattle the surface should be near-dry on application.

6.8.iii Structural block (adobe – American) – temperate climatic conditions

Prepare block material from reconstituted or balanced new mixes with low clay content: 15% maximum clay fraction if nil or low limes are present, rising to 20% maximum if chalk or lime is in the mixture. Include fibrous material – straw, hair, synthetic fibres – in modest proportion if traditionally employed, or required for identification. Accept small pebbles, grits and coarse sands. Mix thoroughly in a soft wet condition and allow to mature in circumstances in which water is withdrawn slowly. This may be achieved by evaporation or by suction into the substrate. The mix may be placed on dry boards covered with sacking, maturing to a stiff mix in 24–48 hours. Remix with mechanical power ultimately pounding into moulds or slab which is then divided. Cut to blocks or units, size to match the voids to be repaired. Allow to stand in good ventilation

in shade in damp atmosphere on absorbent base until hard and near dry. Turn to equalize drying. Do not stand in sun. Bed into place in a wet earth mortar of similar material, from which pebbles and large components have been removed at an early stage. Pre-wet the adjoining areas with light spray.

Blocks may be bedded dry or semi-dry. If dry, the surfaces to be mortared should be lightly wetted immediately before or during placing.

Where weak but stable blocks are required, they may be formed from mixtures in which clean earths are modified with an equal volume of PFA and/or brick dust, hydrated lime and lime putty, and a further equal volume of sand, i.e. 1 earth, 1 PFA/brick dust, 1 hydrated lime, 1 lime putty, 1–3 sand. The sand content may be increased up to three times. The mixing must be thorough and the units should be compacted or compressed, being maintained damp for several days before being slowly dried. The resulting material will be usable as a weak fill block to be used in place of earths which have been cut out due to structural failure or wetting. Samples using the available earths must always be tried and the mixes varied to achieve a material comparable in density and strength to the existing walling. The volume of sand and the extent of compaction will determine the nature of the finished block. The earth component is essential to the initial coherence in preparation. The earth should be free of organic material, dirt and products of organic decomposition. The blocks will be light in colour and are suitable where a surface finish is to be applied.

6.8.iv Stiff earth filler

Repair in a stiff plastic earth (e.g. cob or solid walling) may be undertaken with a stiff mix of just workable consistency placed or pressed into a confined space or behind a shutter. A graded material which will minimize shrinkage is essential. For this purpose, whatever the original mix, a careful formulation eliminating expansive clays is the aim. A substitute material can be a mix of 1 lime putty, 1 hydrated lime, 2–4 PFA/brick dust, 2–4 sand. This is modifiable with native earths of the non-expanding type at the expense of the PFA and some sand. The mix is not matured, since it is effectively a weak mortar.

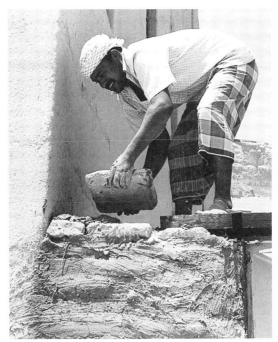

Figure 6.6 Mud brick in manufacture. In arid climates the loss of water into sandy ground equates to loss into the air, and in such favourable circumstances turning and shading are unnecessary. The sandy base permits shrinkage (courtesy of Rebecca Warren).

Figure 6.7 Earth mortar in preparation: the mixture has been prepared wet, including dry straw. It has then been allowed to mature for 24 hours, has been rewetted and remixed prior to use (courtesy of Rebecca Warren).

Figure 6.8 Laying new earth block (mud brick) in earth mortar: the brick is semi-dry and the mortar stiff. An initial keying coat of mortar is applied to the surface. This technique is appropriate to restoration.

It should be well-worked and eased into cavities with a probe and manipulated or compressed with suitable tools. Fibreglass rods or non-ferrous mesh may be introduced as a reinforcement or a tying material to provide a linkage to the existing structure. Cavities should be undercut to achieve a keying. Alternatively, a physical bonding of inserted tile slips may be considered using tile fragments up to 80 mm long – the equivalent of large aggregate – built into pockets in the original material or pins, skewers, proofed wooden pegs and similar mechanical linkages. In conditions where wetting is not anticipated, gypsum plaster may be used in place of or in addition to the limes.

7

Organic materials in consolidation and repair

The invention and development of synthetic organic compounds capable of cohering and binding together particles of earth have attracted innovators whose experimentation is yielding significant, if often negative results. Important work has been carried out by the Getty Conservation Institute at sites in the south-western USA and in China.

7.1 Modifications in the performance of earths

The different effects which may be expected from materials used in organic intervention in earth building can be grouped as follows:

1. Consolidants: a consolidant acts at the near-molecular level by fixing or inhibiting the capacity for movement between very small particles, thereby altering the characteristics of the material in terms of its behaviour, particularly in the presence of water. It tends to make the material stronger in compression and tension, and may affect inherent characteristics such as heat and sound transmission and rigidity.
2. Adhesives: materials causing particles to bind together without necessarily increasing rigidity or stiffness. They necessarily increase tensile strength but may not increase compressive strength. The modulus of elasticity is always amended by the introduction of an adhesive. An adhesive may work in a number of ways: in the most common way, long chain molecules wrap spaghetti-like on the surfaces of and between adjacent particles,

binding themselves and the particles into a more or less coherent mass. There is no clear boundary between the performance of a consolidant and an adhesive.

3. Hydrophobic materials: by their presence in the pores they reduce the attraction of earths to water. These are water repellents. They act by altering the attraction between liquid molecules and the adjoining structure. Hydrophilic materials have the converse effect.
4. Sealants: materials which close off spaces in earths and render them impenetrable, primarily to water.
5. Fillers: materials which provide additional bulk in earths without necessarily rendering them impervious or non-porous and without increasing the binding or bonding between particles. They are likely to increase the compressive strengths of the material.
6. Carriers: materials (usually but not necessarily liquid) in which material to be introduced into earths can be implanted for transport into its final position.
7. Stabilizers: a general term used by earth conservators to indicate one or some of many forms of modification to natural earths which render their particles less likely to move. Most stabilizers work by enhancing the crystalline network and so increase rigidity.

Stabilization of soils involves restructuring:

*   by reducing internal voids through compression of the soil
*   by introducing a material which will glue or cement soil particles together

Figure 7.1 Work of the Getty Insitute at Fort Lauderdale, New Mexico, 1990. Test panels set up in the open to provide data on modifications to earths in weathering, together with control panels, immediately adjacent. An important aspect of this research has been the sustained trialling of wide ranges of materials. For the first time on a large scale and with consistent techniques earth as a basic building material has been exposed to careful monitoring of its behaviour in scientifically prepared and supervised conditions. Only a sample of the trial series can be seen in these pictures.

- by introducing a material which will occupy voids and provide a linkage holding particles in place
- by causing particles to bond chemically or electrostatically by changing their ambient circumstances
- by inhibiting the arrival or passage of water into and through the pores of the soil

7.2 Terminology

In discussion of materials and methods of intervention the following terms are current:

- Organic: a compound essentially containing carbon and hydrogen, usually in association with oxygen

- Chelate: an organic compound incorporating metallic ions (by dative covalent bonding). A chelating agent is the organic compound (loosely, an acid) into which the metallic ion (e.g. calcium + sodium + cations) can be introduced. Ethylene/diamine/tetraacetic acid (EDTA, used in laboratory tests for carbonation) is a chelating agent. An important property of a chelating agent is to render particular elements soluble and available for reaction
- Resin: in other than the natural sense – an organic polymer of long chain molecules capable of plasticity
- Ester: an organic equivalent to a salt formed by reaction between an alcohol and an acid. As a result of the reaction hydroxyl ions become part of the ester

- Thermoplastic and thermosetting: terms generally applicable to resins and indicating materials which, respectively, can pass above softening point and then return unchanged in properties on cooling and those which are permanently changed and made rigid by heating
- Hydrolysis: the engagement of a substance with water as the result of which it is broken down into its component parts. It occurs with some inorganic salts in solution, some esters, non-metallic chlorides and many complex organic compounds
- Polymer: a resin or wax or similar material found in a form in which a number of single organic molecules have become sequentially linked without structural change i.e. the substance remains chemically the same although its physical characteristics may differ. The simple molecule is a monomer; linked, it becomes a polymer. The linkage may be as independent chains or with cross-linkage between chains

7.3 Natural organic materials

Organic materials used for strengthening earth structures are almost invariably liquid or semi-liquid and normally are dispersed in a carrier. Prior to the introduction of the extensive range of resins available through the petrochemical industries, natural oils and resins and various albuminous products were used. With few exceptions the latter do not concern this study, since their function has not been that of a preservative or stabilizer: they have been lubricants giving slip and cohesion in the application stage and probably having some ameliorating function in the preparation of the material during any stage of resting between initial preparation and application. In the longer term such materials have generally disappeared from the earths either by bacterial degeneration or as microfoods. Two persistent exceptions are linseed oil (used in putty, seals and paints) – a compound oil, which hardens by oxygenation in air, derived from the seed of flax. Putty has its own special niche in building technology and has even been used in the fine pointing of brickwork.

Likewise this section is not concerned with natural or synthetic reinforcements – hair, straw, terylene, etc.

The great success story, in one sense, of natural organic stabilizers lies with asphalts and bitumens which have been used for the purpose for a very long period, having proved useful and permanent. While they have the one fundamental defect of irremediable discoloration of soils and the secondary defect of containing viscous long chain molecules, which makes their introduction into earths difficult and in many circumstances impossible, the heavy fraction bitumens share the useful characteristics of being soluble in the lighter hydrocarbons, viscous to the point of near solidity at low temperatures and more fluid at higher ambient temperatures.

Asphalts are essentially heavy fraction bitumens, modified with filler materials such as sand and granular calcium carbonate. Bitumens can be emulsified as extremely fine particles so that they can remain in suspension in suitable aqueous solutions. They can also be carried in various natural and synthetic hydrocarbons. All such carriers can evaporate off, leaving the bitumens in place, and where they have been carried into soils they act as hydrophobes, fillers and consolidants and in this sense can be used as stabilizers. For effective distribution physical mixing with the soils is the effective method. Carrying bitumens into soils in solution or as emulsion is ineffective because the material is self-adhesive and coagulates to fill the pores, inhibiting penetration. In terms of conservation the use of heavy bitumens in earth admixtures is important but limited to damp inhibiting layers which are screened from vision and from which the migration of dark-coloured material does not intrude on other surfaces. It is also used as a surfacing material, where this is historically correct. Bitumens, derived from coal tars rather than crude petroleum, have had traditional application in parts of the northern hemisphere – in East Anglia, UK, for instance, where they have been used in such a way as to create a permanent semi-flexible skin lightly adhering to an earth base and renewable at intervals of 10–50 years. These were originally applied as hot tars, then as cold tars in solvents and currently as water emulsions compatible with the cold tar base.

A further significant application may be in curtilages where extensive water-proofing or water inhibition of ground areas is required. Ground stabilized with crude oil or more sophisticated variants can be surfaced with thin natural earth or other suitable natural materials to provide aprons which are sufficiently waterproof to deal with flash floods, general intermittent rainfall and heavier wear than would be acceptable on normal ground base. As a solution to practical problems they may be recommended in circumstances of low capital investment, minimal plant growth and high wear.

Optimum effects are achievable with moderate volumes of bituminous stabilizer. Little further improvement is obtained by introducing more than 5% asphaltic bitumen and about 3% is usually sufficient to produce a block some 50 times better than a normal earth block under exposure to rain. A particular merit of bituminous compounds is their resistance to biological degradation – a feature not shared by natural oils and resins of vegetable origin, many of which have been tried as soil stabilizers: some have had a degree of success but none have shown good evidence of permanence in the face of biological attack. Many also fail under exposure to ultraviolet light.

It is not possible to produce a definitive list of materials tried but those which have achieved notice often include oils which oxidize to produce tough weather-resistant finishes such as linseed oil, and ranges of resins suspended in plant juices as emulsions, such as the juice of the agave cactus, a material traditionally used in Peru and other parts of South America.

Oils of vegetable origin have achieved no place in earth conservation other than as binders for putties and paints and the conservator is unlikely to consider them even for experimental purposes except where they are a traditional material. Of the resins incorporated in plant juices it can be said that their practical use has been established in limited vernacular circumstances, principally as hardeners in the surface finishing and polishing of earth structures. Such resins are obtainable by boiling stems and pulpy sections of plant materials, including some varieties of species such as *Ficus, Euphorbia, Eucalyptus,*

banana and cactus. Some other plants producing gummy substances may have similar potential.

Many other natural resins have been used in locations related to their availability. Copal, as Manilla copal, has a water-repellent effect, as does Wallaba resin in lateritic soils. Shellac, gum arabic and rosin (tree products derived without metamorphosis) all confer improved strength on soils and reduce water penetration. They are generally carried by an organic liquid but are all water-soluble to a degree and therefore of limited use in exposed conditions.

Experiments carried out at the Catholic University of Peru and field trials of other materials at Chan Chan in Peru have indicated successful stabilization by the use of natural resins, but evidence of the longevity of the treatments is uncertain. Species from which test results are available include: *Ceratonia siliqua* (locust bean), *Euphorbia lactea, Musa paradisiaca* (banana) and *Opuntia ficus indica* (cactus).

Tests on the cactus material showed that the extract was dependent for efficacy on careful control of the temperature and timing of a cactus water-leaching process. A soaking time of 18 days has produced a material which, when used in the construction of an artificial (i.e. non-historic) soil block, yielded a material with a resistance to erosion equivalent to that of a block stabilized with approximately 3% asphalt by dry weight. The permanence of this treatment is at best uncertain.

The relevance of these and other studies is not so much in the availability of the material itself but in the circumstance that the conservator chooses to use traditional methods in conservation operations and seeks to apply modified native lore to the repair of vernacular buildings.

Some materials used for consolidation or stabilization are derived from natural bases by more sophisticated processes than direct extraction but can hardly be called synthetic. Foremost among these are:

- Lignia, a byproduct of the paper-manufacturing industry. It is resinous and resistant to bacterial breakdown but water soluble and therefore of limited value
- Vinsol, a byproduct of turpentine distillation which is water-repellent and has some

consolidant properties but fails in alkaline conditions. The application rates are sensitive

- Aldehydes deriving from molasses which are highly viscous but capable of admixture with dried powdered soils to produce coherent earths. Other aldehydes include furfural which is a product of seed structures from which derivatives using aniline and resorcinol can be constructed. They are stable binders in earth and also act as water repellents. They are generally toxic and bacteria-resistant

7.4 Synthetic resins

The methodology and application of synthetic resins have been much more carefully studied and documented. Despite this, no researcher would claim that the use of synthetic resins in conservation of earth structures is at other than an early stage of study. Many of the investigations are being conducted in investigative programmes in simulated circumstances rather than on historic fabrics for the good reason that failed experiments can sometimes be much more damaging than natural erosion. Except in emergency the wise conservator will not commit significant historic fabric to the burden of uncertainty and experiment.

Simulated conditions may well vary from field circumstances on historic buildings. It is particularly evident that reconstituted earths do not behave in the same way as earths which have been in a structure for a long time and it can only be concluded that the evaluation of the use of synthetic organic chemicals in the conservation of earth structures has some promising aspects and that evaluation of such promise will be an extended but increasingly rewarding process, studded with failures. An exemplary process of investigation involving several materials examined by X-ray diffraction analysis, thin-section analysis under polarizing light, chemical and atomic absorption analysis and physical testing on sample cubes concluded, in the case of samples from Abu Sir, Egypt, that tetraethoxysilane was the most suitable consolidant in that particular instance.

The ideal material will confer stability to the earth structure, giving increased resistance to water penetration and abrasion by wind-carried particles. It should be water-repellent but not so hydrophobic as to create a transition zone. It should leave the pores of the material unaffected so that permeability is not decreased. It should be capable of being carried into earths, leaving a progressively lower concentration to avoid forming a hard transition layer. The necessary stabilization should be achieved without a high volume of introduced material and it should have the general characteristic of reversibility, avoiding significant change of colour.

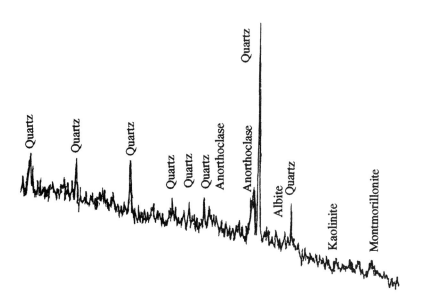

Figure 7.2 X-ray diffraction from a sample of ancient mud brick deriving from Abu-Sir, Egypt.

The varying patterns of diffraction of X-rays recorded as the scanner moves across the range of wavelengths provides an identification of mineral types. It is a qualitative, not a quantative analysis, revealing mineral structures as opposed to elements.

A particular difficulty in the use of synthetic resins is that the process of polymerization which confers on them many of their useful characteristics also makes their molecules sufficiently large to affect their movement through the smallest pores of compacted earths. For this reason an ideal material will be introduced in its minimal condition – as a monomer – allowing the components to link together to form the polymer in their final positions. In some circumstances linkages tend to form 'ropes' whose binding effect produces the necessary combination of stabilization and elasticity. In other cases the link is simply a bridge. These products are characterized by large molecule size and in some circumstances the molecule size is so great that penetration of minute pores in the surfaces of the clay plates is inhibited. A monomer which has been polymerized *in situ* is methyl methacrylate polymerized by benzol peroxide. Other polymers in use are polyvinyl acetal, polyvinyl acetate and polybutyl methacrylate. Various commercially available emulsions are mixtures of similar polymers. The viscosity of some materials precludes their introduction by a carrier, restricting their use to physical mixing and therefore the reconstitution of the earths. Others may be synthesized *in situ* with the appropriate catalysts. This is particularly true of hydrolysis reactions where the moisture present within the soil is the reagent.

The perfect material has not been found and probably does not exist.

7.5 The use of solvents

While water is a powerful solvent and a useful carrier of emulsified material, it suffers from the prime disadvantage of causing swelling in clays by its penetration of the interlayers. This is particularly true of clays containing a high proportion of montmorillonite (the smectite group). For this reason non-aqueous carriers from the hydrocarbon ranges – mineral spirits and the paraffins – have attracted innovators. Other more volatile solvents have included the alcohols, acetone and xylene/toluene mixtures. Many solvents are in fact almost unavoidably mixtures. Ethyl alcohol, for instance, is not readily available without the admixture of water.

Behaviour of solvents and their mixtures is in itself a specialist study. In many instances the conservator must accept that a carrier that is commercially available and economic will be a mixture. The carried material must therefore be tolerant of all the components in the solvent mixture. The carrier itself may necessarily be a mixture in order to dissolve the materials to be carried and these themselves may also be complex mixtures. It is immediately obvious that the permutations between mixtures of consolidants and mixtures of carriers are extensive, with variations of solubility demanding complex introductory procedures to achieve the desired solution.

When the variations in soil types and the variations in density of solutions are added to these factors, it can be seen that a wide variety of permutations is possible. The conservator is therefore left with the choice of accepting the limited ranges of proprietary materials or seeking specialist advice on the formulation of special compounds and application techniques. While specialist advice is obtainable, it tends to be derived from sources with experience in specific limited areas. There is no current professional or institutional source base for advice across the entire spectrum of earths and materials available for their consolidation. The common method of procedure is therefore to seek advice from those experienced in comparable problems – a pattern likely to persist for some time while conservation institutes and agencies gather and build upon experience.

Intermixtures of technique provide one of the most promising and sophisticated fields of potential, loosely describable as composite structures. Thus a soil which might be ineffectually restrained by a particular geotextile might behave entirely differently if even moderately modified by a consolidant and the combination of the two materials may be more efficient in cost and in longevity than a heavier and more expensive application of either alone.

Solid consolidants in the form of fibres are part of the historic technology of earths. Synthetic materials, with their greater strengths, resistance to decay and availability as mesh or fabric offer advantages and new methods of achieving sensitive solutions to difficult problems which will contribute importantly to the conservator's armoury. The inclusion of rods and fibres is to be seen in general terms as earth reinforcement.

To this point the inorganic fillers and consolidants considered have almost all been materials delivered in solid form as powders. Powders and small-scale solids may be carried in agitated liquid suspension, or dispersed in a plastic mass in which a liquid, usually water, acts as a lubricant. Smaller insoluble particles may be carried as an emulsion. Such particles often derive from immiscible liquids such as resins which are emulsified in water. The essential principle of emulsification is the subdivision into particles so small that they are susceptible to molecular agitation. Some emulsifiable particles may be generated from solution by precipitation as a reaction takes place chemically or physical conditions alter. Such alteration can include changes in temperature, in the mixture of different liquids acting as carriers, density of solute, pH value and concentration.

The objective with all these materials is to place the organic compound in a position where it can be used as a bond and strut between the plates and this can only be achieved by dispersion of the compound within the earth or clay. This dispersion is almost inevitably carried out by placing the compound in position in solution; the solvent ultimately evaporates and leaves the compound, or a conversion of it, in position. It may also be desirable – and indeed is almost inevitably desirable – that the density with which the compound is distributed through the earth is greatest at the outside, decreasing gradually, perhaps for a depth of up to 300 mm. The behaviour of the solvent which is the carrier is therefore very important and if several compounds are to be placed the carrier must be suitable for use with all of them. Carriers which have been used are the lighter hydrocarbons including the paraffins and petroleum; carbon tetrachloride; ethylene dichloride; trichloroethylene; acetone; ethanol; toluene; xylene; isopropyl alcohol; ethyl alcohol; and they may incorporate water, which is frequently used to carry a compound in a solution of some other carrier. For example, a compound soluble in alcohol but not soluble in water may, in some circumstances, be soluble in water if it has first been dissolved in the alcohol.

With the increasing use of organic binders the range of carriers required is increased. Many organic solvents are miscible with water and with each other but some are immiscible and some will mix only under certain conditions such as presolution in a third liquid or at specific temperatures, dilutions, etc. Each has its merits in terms of dissolving particular substances such as resins. Volatility, viscosity and cost are other key factors in the choice of a carrier.

The behaviour of solvents will influence decision-making and sometimes lead to compromise. An expensive solvent may with advantage be diluted with a less expensive material once the material to be carried has been dissolved. In other instances solution can be achieved only by the use of one type of solvent which need not ultimately be the prime carrier. The nature and condition of water are notoriously variable; purity, polarity, ionization and temperature all control its capacity to dissolve materials and their behaviour within it. Resins can form gels and flocculants which affect their usefulness in solution. Further problems arise with the presence or absence of catalysts which may be necessary to initiate or accelerate reactions but which in turn bring with them problems of solubility. On the completion of their useful participation in the processes they may themselves become pollutants or agents of decay.

Further problems arise with the application of materials. Once a monomer has been discharged into the earth and has polymerized, the function of the carrier and the catalyst will have ceased. Both may vanish by evaporation, becoming pollutants in the air. They may, however, engage in further reactions in the material or they, or some part of them, may survive for a long time ultimately to produce unwanted products which may be deleterious on a time scale beyond the usefulness of the deposited polymer even if they do nothing more than act as attractants to biological agents of decay. The physical behaviour of the carrier in its disappearance can carry with it further problems of reverse migration, in which a successfully deposited material is then carried back some distance to form an unduly high concentration and, thereby, induces

Poiseuille's law, which describes the movement of water through permeable structures, has been adapted (Taylor) to allow the calculation of flow of liquids; this is a matter of importance where permeants are used as carriers.

Void ratio against permeability   Permeability of kaolinite

The modified formula is $k = D_s^2 \dfrac{\gamma}{\mu} \dfrac{e^{\tilde{}}}{(1 + e)} C$ where k = the Darcy coefficient of permeability
D = some effective particle diameter
γ = unit weight of permeant
μ = viscosity of permeant
e = void ratio
C = shape factor

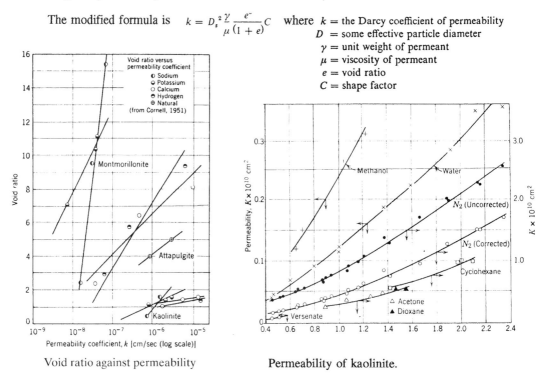

Void ratio against permeability   **Permeability of kaolinite.**

Hence it becomes possible to calculate the performance of fluids as carriers in clay soils. The consolidant carried has a critical effect on the viscosity of the permeant (the combined carrier). Viscosity, polarity and unit weight are critical factors in permeants – in soils, void ratio, particle size, fabric, composition and degree of saturation.

From the left-hand diagram it will be clear that kaolinite clays are more permeable (by a factor of as much as 100) than potassium and sodium montmorillonites. From the right-hand diagram can be understood the comparisons which may be made between carriers, N_2 being a test liquid.

Figure 7.3   Permeability in earths: the rate of movement of liquids through soil.

abscission on an internal layer, with the result that what might have been a successfully consolidated zone becomes a layer of hardened material which peels off. The relationship of all these introduced materials to soluble salts taken up by earths in the course of their natural existence will vary through tolerance levels to failure. Quite sharp boundaries between successful tolerance and failure will produce zones of disco-ordination in the base material.

All these phenomena must be allowed for in designing the ideal method of application

to the base earth and in the final test this can only be done in the field, perhaps on the monument itself. Careful localized testing will therefore be necessary, often over a very long period. Inevitably a considerable time lag must be allowed for making localized experiments and evaluating the results. Where important material is at risk other forms of temporary protection may be required, particularly where non-reversible treatment may be contemplated on a large scale. In preparing for experiments of this nature careful soil analysis is essential. It is important to know not only porosity, grain size distribution and basic performance characteristics but the clay types in proportion (e.g. montmorillonite is highly reactive to water, kaolinite is less swelling, etc.) and the basicity or acidity as well as the presence of calcite, iron compounds and other staining materials.

Among the carriers, which may be carefully selected mixtures, the choice will be made between water and organic solvents, among them the lighter petroleum hydrocarbons (benzene, the paraffins and petroleum mixtures), acetone, xylene, toluene, carbon tetrachloride, ethylene dichloride, ethanol, trichloral ethylene, ethyl alcohol and isopropyl alcohol. Not all these materials are intermiscible; some are inter-reactive chemically and can break down the useful properties of the solutes or even produce hazardous exothermal reactions (to the point of self-ignition). Catalysts can affect the reactions. No admixtures should be made without knowledge of the reaction consequences, and where such knowledge is lacking, careful and safe testing procedures must be used. Many of these products are flammable, some are hazardous within the meaning of statutory enactments in different countries, some are or can produce poisons, some are carcinogens and many are harmful if misused or handled without care. Full precautions should be taken in their storage, handling and use. The effectiveness of carriers can be influenced by their polarity and by the weight of base material in solution. Some materials and solvents will be supplied unmixed for reasons of safety or because reactions begin on mixing. Unless the conservator is a chemist of considerable experience in this particular field he or she is unlikely to build up sufficient specialized knowledge to create formulations but should hope to acquire working skills with supplied formulations relating to the limits and potential of the proprietary manufactured materials that are or will become available. Marketing and legislation dictate the availability of proprietary materials in many countries.

Where emulsions are to be deployed the carrier is almost invariably water. In water emulsification of synthetic resins may be achieved by mechanical breakdown and diffusion and this process is eased by the use of surfactants. Such substances, however, not only alter the behavioural characteristics of the water; their subsequent behaviour may include breaking down the resins by oxidation or by amendment of the long chain polymers. The reduction in molecular length can be significant in terms of the usefulness of the resin even where its chemical nature is unaltered. Long-term effects may also modify the performance of the materials.

7.6 Resin selection

In selecting resins care should be taken to avoid materials which will break down. Failure is usually attributable to ageing, exposure to biological attack, solvents and the effects of ultraviolet light. This is turn requires selective care in understanding ambient conditions. An important qualification to every comment on usage is, therefore, the circumstance in which the material is required to perform and it must be remembered that no general statement in relation to any complex organic compound can be fully meaningful or reliable without careful cross-referencing to all the background conditions of application and comparability with field trials.

Materials used primarily or largely for their water-repellent properties are the silicone group – silanes, such as methyl tri-methoxy-silane, methyl-tri-ethoxy-silane and methyl tri-tetra-ethoxy-silane, used in mixture with other synthetic resins. They have, in their own right, adhesive properties producing cross-linkages and bonds between particles. The hydrophobic properties which they variously possess diminish water penetration of the earths with consequent protection against the movements of clays in hydration and dehydration. They do not, however, fill pores and therefore do not prevent the movement of water vapour which is taken up within clays in accordance with availability.

The range of synthetic resins which can generally be categorized as adhesives and are known in some groups loosely as latexes have been extensively used with some success in the consolidation of earths. The term *latex* was originally used to indicate that a material has rubber-like properties. A number of synthetic materials categorized as latex however have some properties (such as rigidity) which are very unrubberlike. They include polyolefins, ethyl and methyl acrylates, the butyl acrylates and butyl methacrylates, polyethoxyl methanol, polyvinyl alcohol, polyvinyl acetates and polybutyl methacrylates. In various isomeric forms these and similar compounds have differing properties whose relative merits in the structural conservation of earths are at an early stage of evaluation. They commonly polymerize into chain molecules of considerable length and offer physical bonding coupled with polar linkages. Cross-linking in the acrylates, for example, makes for a rigidity not typical of a latex. The interactive behaviour of these materials in mixtures used in earths is not yet adequately understood.

Widely supplied under various trade names for masonry bonding as mixtures containing up to 40% active polymeric solids, these materials are sufficiently readily available and of established performance to be used in trials and application. Generally their use is not reversible by the application of solvents or other simple methods of removal. A further group which has been tried with limited success is the range of polymers of diamine–soluble nylon. This range of tough polymerizable materials may have a potential yet to be adequately explored.

By contrast, quite extensive investigation has been given to the use of isocyanates with promising results. Isocyanates are more likely than the preceding groups to be irreversible. The principal materials reported on have been hexamethylene diisocyanate, diphenyl methane diisocynate and dicyclohexyl methane diisocyanate. These materials, capable of forming urethane and urea linkages and of being polymerized *in situ*, usually in the presence of a catalyst, are the closest approach in entirely organic compounds to the simple consolidant. Synthetic earth blocks and historic earth structures have been treated and the results analysed over sustained periods. Accelerated weather testing has been used to provide comparative results differentiating various forms of application, concentration of materials and of material types. In general terms isocyanates appear to be capable of significantly increasing the strength and shear resistance of structural earths with consequent improvements in abrasion resistance and resistance to degradation in the face of the application of water.

Isocyanates appear to create a more rigid earth structure than the comparable silanes. This may be a product of consolidation effects with isocyanates as opposed to adhesive qualities with the silanes, but published research is speculative on these factors. It does appear that earths containing smectite and montmorillonite alone respond more satisfactorily to isocyanates than the less expansive earths containing kaolinite clays which are evidently more responsive to the use of silanes. Such general conclusions, however, are hung about with caveats over the nature of materials used, the nature of the base material, dilutions, carriers and the whole subject of application and methods.

7.7 Methods of application

Where synthetic organic compounds are suspended or dissolved in liquid carriers there are essentially four methods of application. In the first the material and its carrier are introduced to and distributed in the earth by physical agitation. This is a process which must destroy the historic structural formation, even though it may reuse the historic structural material. In such a process a plastic or fluid earth is produced. Commonly this is a slurry suitable for application as a render or a grout. It may also be used as a mortar. While this is an entirely traditional method of using the material, it can only be justified in the event of the degradation of the original earth structure to the point when it has become amorphous or is no longer a historic artefact. Detritus at the foot of an eroded wall would be a case in point.

The second is by introduction as an emulsion in water. The liquid penetrates by capillary action without deforming the earth. Its depth of penetration is thereby limited and there is a danger of stratification – an impregnated layer is formed with a zone of weakness at the junction with the untreated material. If the treatments are dense the thermal and humidity responses of the two zones will differ sufficiently for an abscission layer to develop. Water vapour pressures may build up behind the impregnated layer with similar results.

The third is placement in an organic carrier. Like water, the carrier evaporates into the

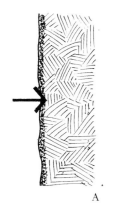 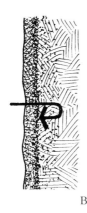 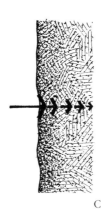

Figure 7.4 Deposition of consolidant by carrier in soil.

A   B   C

A   shallow
possibly due to immediate or over-rapid polymerization or lack of penetration due to pore size/molecular size ratio: result – dense surface layer: failure by separation of the outer skin.
B   shallow
possibly due to consolidant being carried out by evaporating carrier, having achieved adequate initial penetration: result – dense abscission layer: failure by separation of outer layer.
C   adequate depth
diminishing deposition, appropriate balance between polymerization rate of setting time and evaporation or dispersal or carrier: result – graded consolidation layer remaining coherent with background.

Varying effect of relationship between permeant (carrier), rate of setting and deposition of consolidant and nature of soil.

atmosphere leaving the consolidant in position. The density and depth of deposition are determined by numbers of applications, penetration of carrier and relationship of molecular size to pores.

The fourth method of placement involves injection. This technique is only applicable to the introduction of adhesives or fillers to be introduced into zones of detachment or significant areas of weakness. A plaster-enriched earth surface being carried away from its substrate would be a good case in point. The introduction of a gap-filling adhesive in a void or a consolidant-adhesive into a zone of weakness might be justified. Essentially this is a museum technique. Something of a half-way condition is the introduction, by any method, of material which will be intentionally amended and is designed to change within the earth structure. This may occur as the result of oxygenation, reaction with water, ultraviolet light or simply ageing. Quite possibly techniques will be developed which rely on gaseous introduction: the chemical change in lime putty by the absorption of carbon dioxide is a process of this type, albeit natural. The

principal synthetic example is the decay of ethyl silicates (tetraethoxysilane) which release the ethyl component to provide the silicate radical deep in the soil structure to form inorganic deposits which have a cementing and stabilizing effect upon the soil. Some considerable experience has now been gained in the use of ethyl silicate in the field, with encouraging results. The reaction takes place by hydrolysis, causing the formation of ethyl alcohol which evaporates. It also occurs by polymerization, with the formation of water. The silicon attaches to a hydroxyl radical, ultimately bonding powerfully to available oxygen in spaces between and within the clay platelets. It moves from a gel condition to a rigid silica bond which stabilizes the clays, apparently on a permanent basis. As this stabilization does not depend upon long chain organic molecules, and as the porosity of the material is maintained, the technique appears to offer a promising way forward.

Other similar methods of placing inorganic materials in molecular association with soil minerals by chemical change or varied bondings *in situ* may become important, and should be categorized as semi-organic methods.

The use of organic materials as adhesives, consolidants and water-proofers is attended by problems of distribution within soils, decay of additive, discoloration (darkening) and unevenness in application. In addition performances vary widely, depending in considerable part on soil type, climate and utilization. Good general rules are specialist advice, prudence, maximum dilution and careful testing.

A reconstituted earth containing a synthetic stabilizer will at some physical point meet the historic earth structure. At that point the synthetic stabilizer forms a boundary layer with the untreated historic material. The question then arises as to whether some other form of treatment should be introduced into the historic material to provide a less divisible boundary layer – in other words, whether an amelioration of the historic material with the same consolidant or a compatible stabilizer should then be considered. If so, how far should the process go? The ideal conclusion will always be that no artificial boundary layer is created unless it is the intention that the newly superimposed layer is to be readily removable and replaceable. If so, a slurry applied as a surface rendering, with the objective of it lasting 50 years rather than five, might deliberately be designed to pull away from the wall sufficiently coherently to leave the the historic base complete when required. This indeed would be one interpretation of reversibility. Alternatively the conservator might argue that the best long-term protection will be given to the monument if the stabilizing material is distributed in consistently diminishing measure through the entire fabric, producing a better, more reliable long-term bond.

Where the synthetic stabilizer in solution is not physically mixed with the historic earth it will be applied to the historic structure by injection, brush application or spray.

Injection is particularly applicable to materials already wholly or partly polymerized at the point of application, since surface treatments will generally result in a high density of application at the surface with the consequent formation of a skin which will peel away. It is also relevant to hydrophobic components used to inhibit moisture rise of penetration, in effect producing an injected damp-proof course. Injection can deliver calculated amounts of material at specific depths within a structure and by knowledge of the absorptive capacity of the earths the ultimate distribution can be calculated.

The difference between brush and spray application is self-evident; individual circumstances favour one or other technique. More important is the process of multiple coating. Tests on samples will show depths of penetration which may vary between 10 and 100 mm or more depending upon ambient conditions, the nature of applied material, porosity of soil, dilution and carrier. The common characteristic is that depth of penetration is finite and density of deposited material is dependent upon the amount delivered and carried in. Costs of labour, availability of equipment and volume of carrier to be used will all be factors in the calculation of method and performance. Multiple applications may be more expensive than single applications by virtue of labour and costs of carrier for any given amount of deposited material but the results may vary significantly in depth and distribution of penetration. Multiple application, as by brush, usually follows an exponential law of diminishing amounts introduced at each treatment, particularly if reactions occur which progressively fill and seal the interconnecting pore channels. In the long term the consequences will be shown in the performance of the material. A doubling of costs at the application stage may be rewarded by a quadrupling in the intervals between maintenance. However, the law of diminishing returns can also apply and even come into effect in reverse: excessive application and expenditure may produce less than optimum results. Aware of these possibilities, the conservator must formulate test runs and compare the results with other known or similar examples before deciding on a programme of application. Without specialist knowledge of organic chemistry he or she is reliant on specialist advice and the reports of studies. Even the terminology is unhelpful. Chemical formulae specifying atomic components of complex organic molecules are unsatisfactory simply because they collect together large numbers of similar molecules or radicals without describing the molecular construction. One such formula can represent different substances with widely varying properties. Even a single molecule (monomer) of a substance can behave differently from exactly the same material linked, as so many organic compounds can be, into long chains.

7.8 Nomenclature

While it is meaningful to describe water as H_2O – or better, H–OH – a similar collective formula for a methyl methacrylate is ineffectual as a description of the properties of the material, failing to identify isometric variants and polymers for what they are. Even a visible construct of the larger molecular structures becomes so complex that it is not immediately helpful as a pictogram, while verbal generic descriptions can be made precise only with tongue-twisting nomenclature which is by no means consistent in the literature of the subject. For the conservator's purposes, however, such descriptions serve better than most, allowing the materials to be grouped into families and describing polar states sufficiently to provide an accord between manufacturer and user. Proprietary names, acronyms and mnemonics likewise fail to solve the problem. Many names are chosen for marketing purposes and vary from manufacturer to manufacturer as well as from country to country. Furthermore they lack the permanence of chemical formulae: the Creator does not take his products off the market!

7.9 Specification

A chemical description of even the simplest of clays, kaolinite $(Al_2Si_2O_5(OH)_4)$ pales in usefulness beside the simple geological term to which can be added a description of its crystalline state. Thus it is that the geological terminology tends to be more use to the conservator than the chemical in constructing specifications. Thus a 40% solution of butyl acrylate in an aromatic naphtha, short-handed as the manufacturer's description acryloid F–10(R), may be described as being applied in a 10% solution of xylene and the specifier can

then go on to be precise regarding methods of application, temperature, acidity/basicity, mixing methods and methods of application, all of which form further essential parts of the specification. Performance requirements may then be laid down as to the extent of distribution of the material, avoidance of build-ups forming impermeable layers and maximum densities of consolidants. Preparation of surfaces, application of appropriate compatible bonding agents and the availability of materials will be key issues in designing specifications. It should be a general rule that organic solids are deposited in the minimum effective weight. Where the investment exceeds 1% by dry weight of the soil material the conservator must seriously question the wisdom of such levels of commitment. Ideally introductions of consolidant materials at less than 0.01% by weight should be the objective. Stabilization, consolidation and water protection should be undertaken on the basis that the treatment is not to be a modification of the soil structure but an amelioration of its disadvantageous characteristics. Only in the case of special objects in a museum-type context is the conservator likely to go beyond this general rule.

Applications in which large volumes of synthetic resins are applied have not been reported as giving satisfactory results whether as interstitial material or as surface coatings. Urea and phenol formaldehydes have been tried in products designed to provide a durable and impermeable external skin. Success with any material of this type does not seem possible, however flexible and apparently durable the formulation. Colour change, change of surface texture and finish as well as vapour inhibition run counter to any prospect of useful achievement. Likewise, the use of epoxy resin produces a material whose bond to the native earths is so powerful that the resulting component differentiates itself totally from the remaining structure.

The general characteristics of synthetic resin materials differ significantly from those of earths. In thermal performance particular resin treatments cause the behavioural characteristics of untreated and treated areas to differ by wide margins – a factor of as much as 100% of the basic coefficient – and where the volume of introduced resins rises as high as

(a)   (b)

Figure 7.5 A restoration subsequently coated with a gel (possibly urea–formaldehyde based) as a weathering protection. (a) An unnatural sheen is visible. (b) A patch has peeled away. No such technique has yet proved successful.

0.5%, these differential effects will be damaging if a sharp boundary exists between treated and untreated material. At higher concentrations these effects are significant.

7.10 Synthetic reinforcements

The traditional reinforcement of organic fibres and timber or cane rods has been supplemented by other available materials, principally metals, resin-reinforced glasses and plastics. All these materials can be used as fabrics, filaments, rods or rigid bars; the essential difference between the latter two categories is a designed capacity for performance as a strut as opposed to a tensile member. Reinforcement also functions in shear.

In earth structures the bond between a strong reinforcing material and the main structure of earth must essentially be compressive – that is to say, the transfer of stress from the one material to the other will be by way of earth in compression as opposed to tension. In other stronger materials a reinforcement may be effective if glued into position or even welded. It then transmits stress by tensile strength of the bond at the junction of the materials. This effect is minimal in earths. It follows, therefore, that the compressive strength of earths is a critical factor in transfer of loads and the method of use of the reinforcement must avoid reliance on any situation in which the shear strength or tensile strength of the earth is the critical factor. Thus, in the case of a simple bar embedded in an earth structure an applied thrust on its length will be counteracted by the cross-sectional area of the bar multiplied by the compressive strength of the earth. This is not the ultimate compressive strength of the totally compacted material but the immediate compressive strength prior to deformation of its internal structure. After such deformation, when displaced particles take up voids and increase their collateral bonding, the compressive strength will be much higher but the necessary displacement will have allowed the bar to move, probably causing the reinforcement to be ineffective. If a shear force is applied to the bar it will be resisted by the compression of earth on one side and a small amount of tensile or adhesive strength on the lateral faces limiting the resistance to the length of the bar multiplied by its diameter and the immediate compressive strength of the earth.

In any calculation involving unamended earth structures, adhesion must be ignored as a restraint. In an amended earth, however, an adhesive may provide significant bonds between a reinforcing member and the earth itself, both at their interface and in carrying such stresses into the general earth structure. However, in unamended earths the interaction between the earth and the bar depends on compression in the earth and, therefore, as in concrete but to a much greater extent, a shaping of the bar to provide a turn or a hook on its end, and to a much lesser extent a physical roughening of its surface will achieve greatly improved transmission of stress from reinforcement to the basic earth simply because the reinforcement then presents a larger surface against which compression in the earth can act. Seen in terms of a simple bar or hooked bar embedded in earth this argument is unduly simplistic. However, when it is applied to fabrics of any material embedded in earth the effect is multiplied many times and the argument takes on structural significance. If instead of a bar the bonding material is an expanded metal sheet the bonding strengths achieved are important, and the argument can be extended into the use of materials such as geotextiles, which are effec-tive in providing additional strength to earth structures.

The argument can be extended one stage further to describe the interactive effects of random reinforcements. The interlocking of flexible members in sufficient numbers to be in close association determines the ultimate strength. The phenomenon moves out of the tensile strength of fibres alone into the coherence of the granular and stranded structure acting as a unity. It lies behind the efficacy of chopped fibre reinforcement, distributed randomly or in general alignment in earth materials, a technique long practised in the introduction of straw in mud brick and hair in plaster and now extensible to synthetic materials.

7.11 Ideal materials

Earth structures have always been built on the presumption that they will be short-lived or regularly maintained. The fond dream of conservators is of finding a material which is cheap and can easily be applied, and which is capable of solidifying the buildings once and for all, avoiding the need for future maintenance and at the same time being capable of entire reversal should the need arise. While all conservators will admit that the dream is unattainable, it hangs in the imagination like a Holy Grail, drawing them on to experiment further. Many such experiments are crowned with partial success and there are real prospects of limited achievement which will significantly lengthen the useful life span of earth structures by making them more resistant to weathering, physically stronger and generally more durable, reducing the maintenance burden and enhancing their utility.

The ideal material will be reversible, applicable to a wide range of earth types, deeply penetrating, non-toxic and easy to handle and to apply, non-coagulating, water-repellent and will cause no discoloration, surface sheen or other disfigurement. Ideally, it will also be a repellent to insect and biological attack. It will be capable of acting as a glue, as an inert gap filler and as a molecular consolidant. It will be permanent, readily available, indestructible, economic in use and low in prime cost.

If current investigations are any guide, this ideal material does not exist. Nevertheless a number of compounds which fulfil some of these criteria to significantly useful degree have been found, are being developed, are being tested and will become available in the context of appropriate knowledge of their application. Broadly they fall into two groups – organic and inorganic, with a range intermediate between both. With the exceptions of some forms of carbonates they are all virtually synthetic (man-made). Even those which are natural in origin are processed and applied with calculated skill. The inorganic materials depend upon crystalline formations in earths and particularly within clays. The organic materials generally have structures of large molecules, some of which are polymerized *in situ*. Their function is to bind. If, in crude analogy, the inorganics act as a series of little chocks, pillars and struts, the polymers tend to act as chains or ropes. Other organic products affect the hydrophilic properties of soils and some depend for their usefulness on being hydrophobic. Many are dependent upon carriers and catalysts for their successful introduction and will only achieve the desired results under carefully controlled conditions. Some are capable of being used simultaneously in admixtures, some sequentially and some are mutually incompatible. The nature and circumstances in the treatment of soils can have a significant effect on their performance. A soil which may not respond usefully while *in situ* in a historic monument may give an ideal response to consolidation if removed and reworked with the consolidant. Salinity in soils and the nature of weathering can have dramatically different effects upon identical treatments. The types of clay involved may condition the response. Above all, the state of present knowledge is largely inconclusive because most trials and experiments have been carried out in the recent past and, despite some accelerated weathering tests, their certainty and longevity cannot yet be verified.

8

Practical interventions

The conservator will presage any intervention in a major structure with an analysis of the reasons for any action and the nature of the work to be carried out, looking critically at effectiveness, permanence, cyclical maintenance, economy, structural performance, visual effect, identifiability, viability, achievability and the appropriateness of the contemplated options. Many a wrong decision has been made by following first thoughts which have failed to take account of the true nature of the problem and the most beneficial long-term solution. Choices may lie between minimal intervention and recurrent problems as against greater expenditure, more uncertain technical background and irreversible options. Secondary pressures can sway decision-making. An action may become more justifiable because it will provide information for future work or it may (unjustifiably) be aimed at the greater glory of a building owner or a conservator seeking to extend or elaborate the work as a matter of status. These and other influences which affect decision-making call for logical analysis and consistent philosophies.

Generally in earth structures the options lie with various forms of traditional protection, types of filling or replacement and the ever-widening choice of materials for consolidation and reinforcement.

8.1 Protection of roofs and cappings

Earths have been used as a defence against rain in two circumstances – extreme poverty of resources in a climate where there is a damaging level of precipitation and, more commonly, in a climate where rainfall is modest and of short duration. In the first case there is little point in attempting to perpetuate traditional methods without improving the longevity of the material, but in the second the continuity of method may be as important in terms of the local economy and continuity of lifestyle as in the retention of the architectural forms.

In all cases of direct resistance to rain by earths the impermeability depends on the bulking of clays when wetted. The first water penetration causes expansion and this inhibits further rapid penetration. A denser outer surface causes run-off; only a small proportion continues to penetrate. This effect is temporary, but sufficient with intermittent rainfall even when heavy. In sustained wet conditions the roof fails, some of the surfacing is washed away, and initial soak-through turns into runnels – clearly impractical. In the circumstance of intermittent rainfall an earth layer between 100 and 200 mm deep may form a thoroughly practical roof. However, rain penetration through the topmost 15–30 mm produces a sufficiently dense layer to inhibit the rate of moisture movement into the lower layers, which may remain dry for several days even while the outermost surface passes the liquid limit and washes away.

However, this phenomenon is at its most effective only on the first occasion of resistance to wetting because the swelling, which reduces the penetrable voids, thereby inhibiting the flow according to Poiseuille's law, does not return on drying. The affected layer which has expanded has no mechanism causing it to return to the original density, so that on the next occasion the voids are more open and capillarity greater. The surface is then softer, penetration is deeper and the liquid limit is achieved more quickly.

The traditional response to this problem is rolling to compact the surface. This can only be done when the clay-rich material is in a semi-plastic state. The mechanical action of a

Figure 8.1 Raqqa, Syria. The city wall after heavy rain. The sheen on the top surfaces is a reflection from the saturated dense soil which protects the earth immediately beneath from serious water penetration.

relatively light roller is sufficient to restore the density of the surface if the work is correctly timed, so that in semi-arid parts of the world rainfall is followed by a surge of domestic activity on the flat roofs, as the surfaces are smoothed, rolled and the adjacent vertical surfaces pressed or 'wood-floated' to restore their previous densities.

Interventions to overcome these labours are either the provision of a synthetic material in lieu of earth or the inclusion of an impermeable layer. In each case the fundamental profile of the structure can be retained but the colour and texture are lost. In neither case is the solution satisfactory in terms of keeping the quality and integrity of the surface construction; in both it may be justified in giving long-lasting protection to structures which might otherwise be lost. This problem highlights a profound dilemma in conservation of the vernacular, and one which is barely perceived in monumental architecture: how in a society faced with radical changes in lifestyle can labour-intensive and low-grade activities be expected of generations equipped with the education and technology of the 21st century? Whole groups of buildings dependent upon these simple techniques form

the inheritance of nations; some, as in the Yemen, rising to the importance of world heritage status. Where incomes, expectations and lifestyles rise beyond those of the earlier builders a radical change in the detailed technology may be required to retain the quality of the entire architecture, and in this context the labour-intensive use of mud as capping and roofing may be overtaken.

However, the ideal is in continuity of technique that gives continuity of structure. Where this cannot be achieved in social terms the intervention may allow for the construction of an impermeable deck or membrane on which a stabilized earth is reconstituted to the original profile. Lime, cement, bitumen or synthetic organic polymers may be used to provide a more durable and more or less permanent surface. Wearing qualities may be improved while higher rates of run-off will be achieved. The impermeable layer, however, will change some previous characteristics. Thermal transmission may be less important than loss of vapour permeability and the consequent formation of initial condensation within the roof structure. The conservator adopting such a radical technique must calculate the likelihood of this and similar effects having a

Figure 8.2   The mud brick walls of Bokhara, Republic of Uzbekistan. Protection of the head of the city wall has retained battlements and face-rendering while rising ground water held by scree has caused the loss of face material.

dramatically damaging effect upon the structure. A change of regime below the weathering level may for instance change the circumstance sufficiently to provide a habitat for entirely different insect or fungal life whose effect on the load-bearing structure must be considered.

The alternative of an impermeable wearing surface may be considered. A significant change in colour and texture is likely, and because the material will be more highly stressed due to temperature extremes, humidity changes and physical wear, the specification and detailing will be more critical. In some cases the problems may be separated. A tradition in which stone or terracotta wearing surfaces are employed may allow the use of a thin surface membrane beneath slates or tiles. Selection of materials must take into account factors such as the flexibility and movement of traditional forms of construction. Detailing is all-important: the longer-lived the construction, the more so. Edges and junctions may be difficult to achieve in the context of original profiles and decisions, sometimes awkward or painful, may have to be made between practical and pleasing details. Some principles will have to be maintained if the advantages of a new durable material are

not to be lost. There is no point in providing a strong membrane across a wide surface only to see it quickly broken and made useless at a joint where powerful movements occur.

The softness of earth materials and their frequency of repair may have to be supplanted by joints which tolerate movement, such as the upstand and cover-flashing system with under-cloaking as necessary.

If earths are unavoidably replaced on roofs and cappings, significant changes in detailing will be required. The ideal is of course that like is replaced with like, and the earth roofs are maintained in their established form. Improvements such as the selection of improved clays not available to the earlier builders may offer some advantage as a result of wider knowledge and the opportunities offered by modern transport.

8.2 Protection of walls

Three vulnerable zones are recognizable in walls – the head, the vertical surface with its openings and the junction with the ground (its base).

The head of a wall may be protected sacrificially or permanently, or by a combination of both methods. Sacrificial protection provides for the loss of the protective material. A wall, once capped with mud brick or rendered over with a mud–dung plaster, can unimpeachably be restored in the previous manner in the expectation of similar renewal at intervals determinable by the quality of the earths used, the quality of the workmanship and, above all, by climate and erosion. Where the wall profile is a significant aspect of the aesthetic and where in particular other, perhaps better protected, elements of wall survive in their original condition, such a solution is strongly to be recommended.

Where greater protection is desired an earth stabilized with a consolidant (see below), or a modified earth given improved durability by the addition of limes, cements, fly-ash and similar materials, may be chosen. There may be reason to defend the wall against heavier wear than previously or to increase the periodicity of the maintenance cycle. The conservator may choose to differentiate the material newly placed from the existing structure by the inclusion of identifiable fibres. Chopped strands of non-rotting plastics, such as terylene, perhaps coloured, may be used in the first application layer so that at the time of their later exposure it is clear that the remedial work has been eroded.

Incorporation of materials incompatible with earths is to be avoided in other than exceptional circumstances. Such materials include waterproof brick, cements as mortar concrete or as precast slab, impermeable membranes such as polythene and lead, liquid bitumens and asphalts, rigid board materials, inert composites, plywood, timber, steel, etc. Generally the problem in the repair of wall-heads is not so much the exclusion of water as the direction in which it is discharged and any circumstance which causes runnels or rivulets to course down a wall face is inherently dangerous in an earth structure. In a composite structure in which the wall is sheathed (e.g. rendered) the concentration of water behind the sheathing due to inadequate discharge from the top surface is likely to be a rapid cause of structural failure and detachment of the facing material.

Figure 8.3 A wall without capping but on a stone base demonstrating runnels of erosion on typical layered wall construction. A wall in this condition may be fully protected by renewing the rendered face (courtesy of Rebecca Warren).

In temperate climates where high rainfalls are common, earth structures are normally protected by a capping. Brick facings and renderings are common and impervious wall cappings are the rule. Such cappings include thatch, tile, slate, stone and even brick and are always best designed with a weathering detail which throws rainwater clear of the wall surface immediately beneath. Earth structures in such climates always contain a significant amount of moisture as water vapour and interstitial liquid water. This is often sufficient to support the decay of organic fibres and therefore the fixings which have been built in to carry protective roof structures will be suspect and may need replacement. Embedded timbers, which these fixings commonly are,

Figure 8.4 Wu Wai, China. The construction of the compound wall of the Confucian Temple can be seen to be of semi-dry mud brick laid slightly inclined over a mass-earth wall on a stone base which does not protect it against the rise of moisture.

can be replaced with proofed material. Alternatively inert composites capable of receiving screws are now available and can be supplied in dimensions equivalent to fixing timbers to perform a similar function.

Despite the basic rule that all materials should be compatible in such construction, there are circumstances in which reinforced concrete may be considered despite the considerable differences in the nature of the materials and their behaviour under thermal change. Being significantly stronger than the earths, concrete ring beams at the head of a wall can, by weight and solidity, provide a stable element for carrying traditional roof finishes and providing for superimposed loads, such as floors which may be subject to conditions for which the building was not designed. Such concrete additions may also be desirable as earthquake protection if supported appropriately. Supported on nothing but earth they

can be a serious danger. Concrete must never be introduced without the greatest circumspection and always with detailing designed to tolerate and allow for the differential behaviour of concrete and earth.

In climates of moderate rainfall, preformed stabilized earth may be used as a capping material and may have sufficient durability to provide protection to the wall by the formation of a capping course with an overhang. For this purpose it will usually be preformed, densified (i.e. pressed) and possibly bitumen-stabilized. While this may be practical it is inevitably incongruous, since earths in their own right would not be capable of performing in this way. Water discharge from cappings and wall surfaces is a critical factor in their durability and survival. Essential clues may be picked up from the careful study of traditional techniques used in vernacular building and two general principles can be established: first,

Figure 8.5 Runnels on the surface of a mud brick wall showing the effects on different densities of material. The mud mortar, having been laid very wet, has absorbed the run-off more readily, has reached its liquid limit more quickly and has therefore been eroded. Renewal of renderings is all that is needed (courtesy of Rebecca Warren).

Figure 8.6 Mexico. Failure of the cement capping, cracked by thermal movement, has opened the way to the detachment of the rigid sheathing of the mud block core (r.h.s.).

that water is discharged evenly so far as possible and second, that where a concentrated discharge is unavoidable it is controlled by being piped, if appropriate, or thrown clear by gargoyles. Factors of microclimate can also be important. A prevailing wind direction during a wet season may have caused traditional builders to discharge water away from the wind direction where rainfall is persistent, whereas in conditions of intermittent rain discharge may be towards the wind direction when there is sufficient drying interval, perhaps due to diurnal conditions. In some circumstances overriding design criteria make it unavoidable that run-off is concentrated. In a crenellated building embrasures may be designed to concentrate water and shed it, and a stone drip is included for the purpose. Techniques of water channelling on the face of a building using inset stone facings or perhaps timber or metal have occasionally been sufficiently extensive to form a design feature. External chutes for sanitary discharge have been a feature of earth buildings for so long that it is probable that the better-known use of the same device on masonry structures was copied from earth buildings. Such techniques are fragile or carefully judged and their overloading can have disastrous consequences, as was discovered in the Yemen

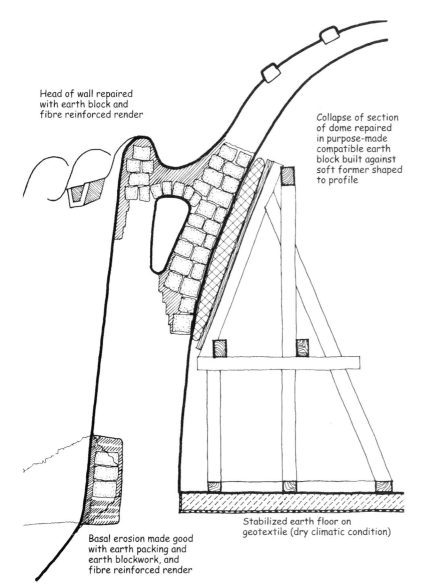

Figure 8.7 Repair of major structural collapse (courtesy of R. Deefholts).

Head of wall repaired
with earth block and
fibre reinforced render

Collapse of section
of dome repaired
in purpose-made
compatible earth
block built against
soft former shaped
to profile

Basal erosion made good
with earth packing and
earth blockwork, and
fibre reinforced render

Stabilized earth floor on
geotextile (dry climatic condition)

when the introduction of modern sanitation and water supplies overwhelmed the capability of traditional drainage systems, causing disastrous collapses in tall buildings. Water running down external vertical gutters in volumes far greater than ever before found its way into the core of the walls, taking earth mortars beyond plastic, or even liquid limits, precipitating an avalanche of rubble and mud.

In earth structures the topmost parts often include parapets to shield usable space on roofs. From flat roofs the discharge of surplus rainwater has traditionally been achieved by gargoyles in the form of gutters carried through the walls, projecting sufficiently to throw water clear of the base. Stone gutters, French drains and soakaways were often provided to gather the run-off. Repair and restitution of these rainwater systems is frequently all that is necessary. The introduction of vertical pipework is usually an alien feature to be avoided.

The vulnerability of earths to water erosion demands that the conservator makes a careful judgement of the capacity of natural systems which may not have been well-designed for their purposes and may indeed have been a prime cause of failure. It may be possible to improve the system of water discharge inobtrusively. The introduction of impermeable sheet materials such as lead, bituminous sheets or plastics introduces the difficulty of obtaining fixings in earth structures. The behaviour of such materials demands slip joints and cover flashings which may for reasons of fixing and unsightliness be impractical. In principle they are undesirable and should be treated on the basis of a presumption against their use. In some circumstances the introduction of a reinforcing material in the wall may be sufficient to provide for the necessary stability and adhesion. Proofed timber members, inert materials or indestructible fabric may be incorporated within the soil structure for this purpose.

Experimental protection of the heads of walls which has yielded satisfactory results has included the provision of stabilized soil using cements or ethyl silicate; consolidated earth blocks; underlayers of earths or earth blocks stabilized with asphalt emulsion and blocks, and earth stabilized with latex. This last technique is sensitive in application and has not yet been proven over a long period. Where a material is used whose colour differs from the basic earth, such as a cement admixture providing a grey bloom and an asphalt mixture providing a darkened colour, aesthetic compatibility may be obtained by over-rendering with the natural earths. The reappearance of the coloured material indicates loss of the sacrificial render coat.

In dealing with rendered wall surfaces the normal maintenance regime for an earth face is a further rendering of the original. No special preparation is required as the eroded surface provides a more than adequate key. In many areas the traditional application of

Figure 8.8   A recently applied flat rendering adjoining an earlier section of render carrying a strong hand-marked pattern. Although this differentiation has the merit of distinguishing periods of work, there will normally be merit in retaining the continuity of pattern already established.

render is by hand – perhaps the most sensitive tool for the purpose. The technique results in a surface texture virtually unobtainable by other means. Surface patterning, modelling and deliberate decorative design are features intrinsic to the form and the tools.

The major problem in the repair of walls is structural. Surface runnels, fractures, collapse due to slump, basal erosion and loss of fabric from other causes require the introduction of new compatible material. Associated problems may have to be solved and these operations may in some cases require the introduction of reinforcement calculated to transfer loads or provide continuity of strength within the capacity of the earth structure itself. The prime difficulty is shrinkage. A lump of plastic earth introduced into a cavity to fill it, pulls away from the adjoining surfaces as it dries. However, a mud render can be applied to the wall because the thin coat of material takes up shrinkage in the depth of the application and the wall surface, drawing moisture from the applied material by capillary action, reaches a compatible state at the interface so that an adequate bond will be formed under pressure of application. Unless precautions are taken this will not happen when earths are introduced into a structural void. Perimeter cracking overcomes adhesion, causing discontinuity and loss of structural support. By careful selection fillers may exclude the expansive clay fractions and the introduced material may then provide the structural support required. The filler, however, will then be different from the original. It may, for instance, be a sulphate-free pulverized fuel ash (PFA)/lime mortar. A problem which is then apparent is visual acceptability.

Where natural earths comparable in performance to the original material are specified and used, the designer must ensure that the shrinkage which occurs will have the minimum effect on the final coherence of the structure. This is probably best achieved in most cases by the use of carefully cut earth blocks built into the void with a minimal jointing which is then filled with a thin mud mortar or slurry of the same or comparable material or, even better, a tamped stiff plastic mix. On occasion a block can be specifically formed to match the void and, where structural cohesion is important, the introduction of fibreglass or

other reinforcing rods positioned to resist shear can be considered. Where additional tensile strength is required synthetic fibre may on occasion be introduced. During the process of introduction of the new material temporary support may be required and this should always be provided using a soft or compressive pad at the point of transfer of load.

8.3 Filling voids

Dried earth blocks may be bonded into damaged walling. In this process physical keying, producing a joint with multiple offsets or undercuts, can be important and can contribute as much to the final strength of the wall as the careful preparation and insertion of materials. To be avoided in such processes of repair are straight vertical joints and thick mud mortar joints, either in perpends or on flat surfaces. The thicker the joint, the more certain it is that a shrinkage crack will exist within the joint and structural coherence will not be achieved.

Ill-considered repairs in earth structures introduce incompatible materials such as concrete in liquid or precast form, rocks, boulders, rubble and aggregate, sand and sandy earth mixes and other formulations which produce unduly hard or soft materials. However, there is a case for considering the use of a dissimilar brick dust and PFA material with properties similar to the earth structure. Success has been claimed for mixtures of fly-ash, brick dust and lime which can produce a setting material free of the problem of shrinkage and with characteristics of thermal movement, strength, resilience, loading and self-weight comparable with an earth structure. These materials should meet the requirements of a synthetic gapfilling component for earth structures and should by their nature have an indefinite life. They conform to the requirements of reversibility (by removal) and identifiability.

The most common error of ignorance is in the application of poured cements and concretes which are incapable of bonding to the earth structure, have different properties and suffer from the disadvantage of hardness and rigidity. In consequence the earths around them fail and the repair becomes isolated and ineffective.

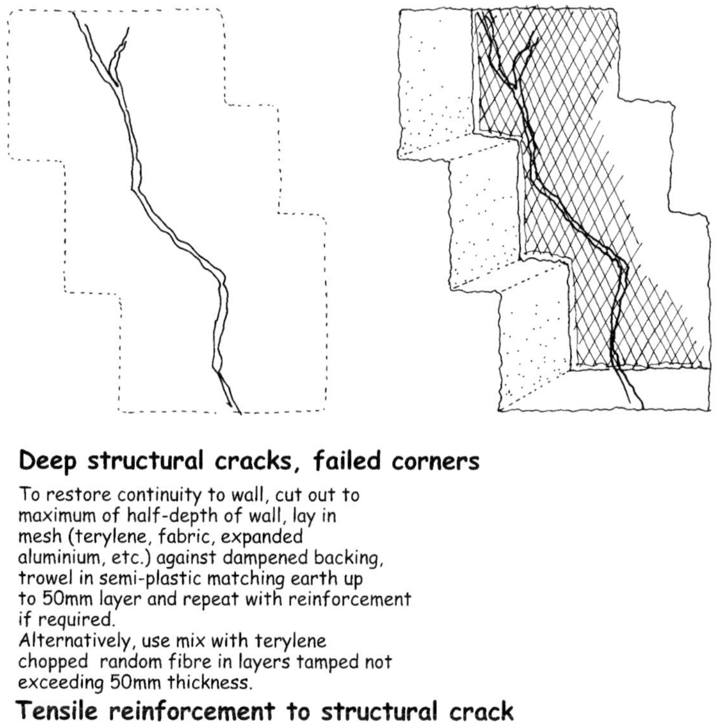

Figure 8.9 (a–b) Deep infill with interlocking mechanisms.

Deep structural cracks, failed corners

To restore continuity to wall, cut out to
maximum of half-depth of wall, lay in
mesh (terylene, fabric, expanded
aluminium, etc.) against dampened backing,
trowel in semi-plastic matching earth up
to 50mm layer and repeat with reinforcement
if required.
Alternatively, use mix with terylene
chopped random fibre in layers tamped not
exceeding 50mm thickness.

Tensile reinforcement to structural crack

(a)

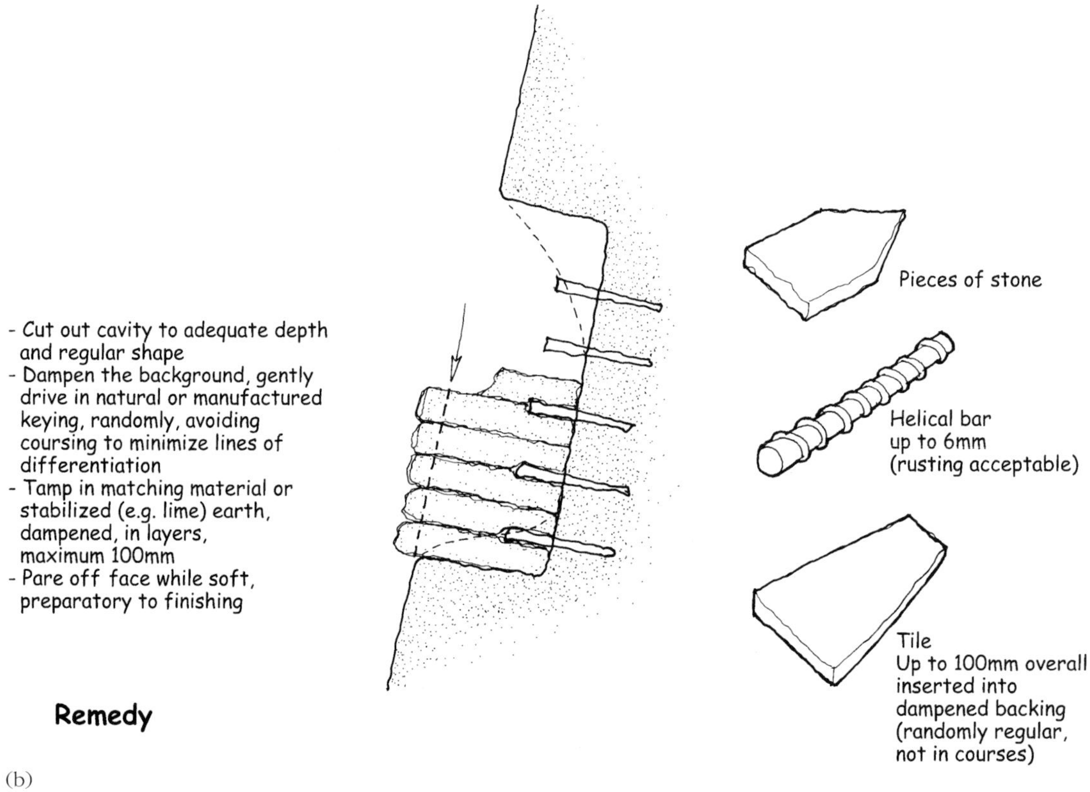

- Cut out cavity to adequate depth
  and regular shape
- Dampen the background, gently
  drive in natural or manufactured
  keying, randomly, avoiding
  coursing to minimize lines of
  differentiation
- Tamp in matching material or
  stabilized (e.g. lime) earth,
  dampened, in layers,
  maximum 100mm
- Pare off face while soft,
  preparatory to finishing

Pieces of stone

Helical bar
up to 6mm
(rusting acceptable)

Tile
Up to 100mm overall
inserted into
dampened backing
(randomly regular,
not in courses)

Remedy

(b)

Figure 8.10 A circumstance where the careful introduction of mud brick components and earth mortar from the lowest level upwards will be capable of restoring the stability of the structure. A liquid fill would almost certainly be disastrous (courtesy of Rebecca Warren).

A common problem of well-intentioned repairs is the introduction of a large volume of clay-rich earth sufficiently wetted to be plastic. This may be pounded into position and a bond achieved with the adjoining structure which may itself have been wetted. As it dries, however, the material inevitably shrinks and draws away leaving substantial cracks to be filled which, having been filled with similar material, again shrink slightly. If, in such cases a mechanical key of sufficient complexity has been formed, the repair may succeed. The technique is traditional and therefore supportable on historical grounds.

Figure 8.11 A circumstance where structural elements of a predominantly earth wall are in contact. The introduction of a plastic fill stage by stage would be considered here (courtesy of Rebecca Warren).

Where earth in a plastic state is to be introduced, shrinkage is much reduced if the mixture is stiff or very stiff. To achieve effective consolidation, it may be possible (particularly at ground level) to provide shuttering placed so as to allow the earth to be tamped in behind the shutter to achieve maximum consolidation.

Where the repair is in a horizontal plane, as in a vault, repairs may be achieved with a material in a plastic or semi-liquid condition. The technique involves the application of shuttering on the soffit or underside, against which a small amount of material is applied in the same consistency as would be used for rendering. Further applications are made in thin layers so that no cracking pattern emerges. Each layer is allowed to approach dryness before the next is applied and, since shrinkage occurs in the thickness of the applied layer, a coherent earth fill is produced. Similar time-consuming methods which require patience may be used in lateral filling.

8.4 Basal erosion

Many problems requiring the insertion of fillers occur and may be at their most

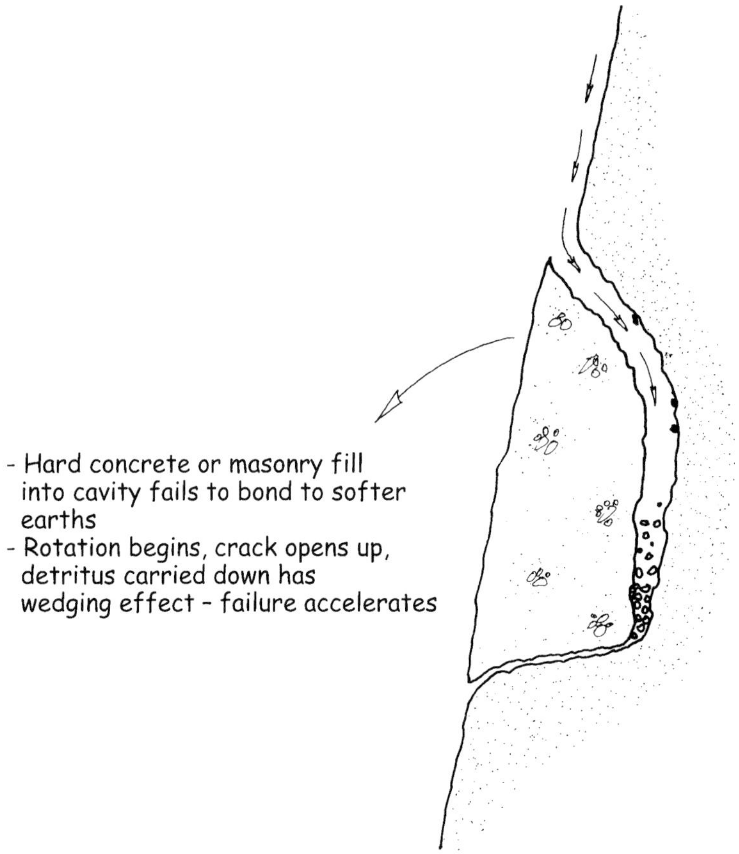

Figure 8.12 Fault intervention (courtesy of R. Deefholts).

- Hard concrete or masonry fill into cavity fails to bond to softer earths
- Rotation begins, crack opens up, detritus carried down has wedging effect – failure accelerates

intractable at the bases of walls. Basal erosion may follow simple weaknesses due to wetting and drying cycles but in many instances they are the consequence of salinity in moisture drawn up through the soil. Chemical analysis will determine whether this is the case. If so, the problem is not simply a matter of replacing the missing material.

Stability of the wall structure will always be a paramount concern and it may be necessary to provide shoring restraint to ensure that no toppling (i.e. rotating) movement is induced. It may be necessary to provide an inhibition against further rising saline moisture and to replace lower sections of the wall entirely to free the structure from salts already deposited. Natural damp barriers include dry-laid rock and slates and these are effective inhibitors which are traditionally used in many areas. If their use is appropriate they may be introduced or renewed in preference to artificial methods of inhibiting rising damp. Impermeable membranes such as are used in brick and masonry structures are inappropriate in earths partly because they can, in some circumstances, trap water on their upper surfaces. In others they produce a sharp boundary between very wet and very dry zones. Their introduction should therefore be avoided. Methods of partial inhibition are preferable. Perforated sheet materials can be applied in some circumstances; in others stabilized earth blocks, tile or stone can be used. Stone of low permeability such as slate may be an effective choice.

The techniques of underpinning applicable to brick and masonry are likewise inappropriate in earths and the practical method of replacing base courses is simply to carry out the removal in short lengths of half a wall thickness at a time. The short lengths should themselves be alternated and generally should

Figure 8.13   The conjunction of a new basalt base and newly rendered earth structure above. A renovation of rather higher calibre than the original and, as a result, somewhat blatantly identifiable (courtesy of Dr Paul Brown).

Figure 8.14   China. Mud brick introduced to combat basal erosion. The different surface texture of the dried brick and the compacted earth above is immediately evident. The effect is entirely compatible. An unrepaired section of wall is beyond. This repair is the work of local labour. Some further packing is required at the head of the repair.

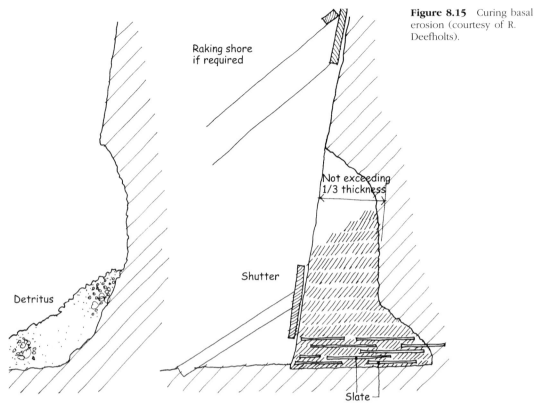

Figure 8.15 Curing basal erosion (courtesy of R. Deefholts).

Raking shore
if required

Not exceeding
1/3 thickness

Shutter

Detritus

Slate

Curing basal erosion and reducing excessive or dangerous water uptake to acceptable levels

- Remove and discard detritus
- Provide shutter and select short sections
  of wall, alternating approximately 1m
  lengths (treated and untouched) inserting
  slate slips in semi-plastic earth mortar
  to build up layer with reduced water paths
- Continue upwards with tamping replacement,
  working slowly to allow progressive drying
- Repeat in untreated lengths
- Repeat opposite face in severe cases

never be more than twice the thickness of the wall. Thus, in practical terms, lengths six times the wall thickness might be allocated for work, each being marked out in three equal sections. If half the thickness of the wall is considered as one unit then in every six units only one should be worked upon at any one time and opposed alternate units would always be dealt with in sequence. An earth structure in sound condition above the level of the repair will normally be self-supporting or require minimum temporary strutting over such spans. Dried earth components introduced above a

moisture-inhibiting layer with thin mortar beds can quickly be introduced. In such instances the shrinkage of mortar can be significant and material with a higher sand and reduced clay content or with illite/kaolinite clay predominance will be selected.

In circumstances where there are good arguments for the use of earth blocks made more durable with consolidants or stabilizing materials, the blocks should be identifiably marked. Asphalt stabilized blocks can be effective as moisture inhibitors and may be considered where they can be concealed.

Adjacent to the earth structure other forms of protection may be sought. The control of water run-off is always important. Soil surfaces tend to move but may be stabilized with geotextiles – non-rotting mesh materials laid beneath the surface. Vegetation may be introduced in addition, to perform a similar function eliminating the need for synthetic materials. Controlled introduction of grass or other plants producing fibrous roots may be sufficient to stabilize soil and to reduce wind-blown particles which would otherwise erode a soil structure.

Buildings frequently have curtilages covered wholly or partially with durable and often impermeable materials – asphalt, paving slabs, concrete slab and the like. The falls to which such materials are laid should be carefully considered in order to shed water away from buildings and the pattern of underground drainage in a curtilage should be considered so that the conservator can be satisfied that all excess moisture is being removed. Even in the most arid climates there can be a danger of flash flooding. Where a cloud delivering a large volume of rain moves along the length of a water course at a speed similar to a run-off the entire delivery of the cloud can be concentrated at one point. Simple precautions such as the availability of a flood discharge channel may make all the difference between the survival of a structure and its total loss.

Special structures may demand the provision of overroofs either carried separately or on the structure, and the provision of wind-screens, perhaps as aero textiles aimed at reducing excessive wind speeds. Structures of this type are invariably obtrusive and their use is likely to be restricted to special circumstances such as archaeological excavations and short-term protection while longer-term measures, such as tree screens, are brought into effect.

8.5 Biological inhibition

Biological inhibition is a subject on which it is impossible to generalize beyond the comment that the fundamental protection is by maintenance. Introduction of fungicides and biocides into earth structures is generally short-lived and may be pollutant or inimical to the coherence of the material. Only in the rare

Figure 8.16 Ar Riyadh, Saudi Arabia. A string-course detail designed to discharge water which has evolved into a decorative device. Also visible are gargoyles designed to throw roof water clear.

circumstance of outbreaks of epidemic proportions should chemical control be necessary.

The essence of biological control is the botanical regime. It is a nonsense to suggest that an earth building should be protected so that it contains no moisture and, therefore, cannot harbour plant growth. The function of a building is to be a shelter, and some structures have from the outset been expected to carry plant growth as part of their natural function. Unless the structure is a bank, a mound or some other earthwork, plant growth is usually a nuisance at best and a danger at worst. Even on an earthwork a tree is likely to be unacceptable and may be kept at bay by techniques as simple as grazing.

On buildings control of moisture is usually a sufficient answer. Ledges and similar impediments to the free flow of water will be avoided, and are an unlikely feature of any earth building. The controlled discharge and disposal of rainwater are vital.

Plant growth on a structure is probably the result of unintended water entry or water retention. A localized pattern of growth will give a clue as to source. A remedy preventing the water entry will usually be sufficient and growth can be left to wither once the fault has been dealt with. Micro plant life may emerge as lichens, mosses and fungi in suitable microclimates. Changes to the local circumstance – wind shielding or removal of shading – can quickly amend the micro-climate and avoid the problem without resort to chemical inhibitors.

In the conservation of earth structures more than any other building type, regular, careful and sensitive maintenance cannot be overemphasized.

8.6 Amendment of earth structures

Historic earth structures should not be called upon to perform structural feats outside the normal capabilities of earth materials because of the risk of visual incongruity. However, there are a number of exceptional circumstances in which physical reinforcement may be desirable, and which the following are examples.

8.6.i Slip and seismic movement

Significant land movement may have taken place, causing or being likely to cause the detachment of part of an earth structure. The introduction of a textile mesh which may also incorporate the additional strength of rods attaching or relating to the mesh may provide a layer of abnormal tensile strength in the base of the structure, causing the whole of the built mass to remain coherent above the base level. The introduction of a slip plane below the reinforced level may be deliberately designed to allow the coherent earth structure to move in relation to part of the ground level, thus avoiding the effects of vertical cracking being introduced into the walls. Such a slip plane

might be introduced by the use of membranes and sacrificial barrier material.

In circumstances of seismic movement a disjunction in the building might be made coherent by the use of appropriate longitudinal rods or bars introduced through the whole length of the structure. These would be rendered integral by plates, pads or textile-reinforced zones in order to adjust the structure to the new loadings and guard against accentuation of the movements by earth tremors. Seismic design may suggest structural severance achievable by introducing vertical cuts at junctions of walls, loss of strength may be compensated for by additional tensile strength introduced by fabric in floors or applied superficially on walls. Earth buildings have a good record of performance under slight shock. The formula of low-rise, high-mass, soft-walled building has inherent advantages in low stress conditions. The advantages are the opposite of those offered by rigid materials such as concrete. To combine rigidity with a flexible structure, therefore, tends to negate the advantages of both by neutralizing them.

Adherence may be given to earth constructions by the use of interlocking earth or stabilized earth blocks. This type of block is designed with an interlocking arrangement and will be a form of adobe construction. The block will have been compressed in a mould and may have been stabilized or reinforced with fibre. As a result it will be stronger than normal and, if not used throughout a structure, it may be introduced as a series of bands or ring beams. The objective is to provide for stress transmission. This type of construction is aimed at new-built work in seismic zones and has only limited application in conservation work.

There may be exceptions to every rule. Composite structures with brick or stone quoins, arches and dressings are but one example. Precautionary seismic engineering may justify the introduction of columns and framework designed to secure the safety of personnel in the event of collapse – a precaution which is particularly applicable to public buildings and parts of homes. Design of such structures is within the competence of only a limited number of engineers and few conservators. It must take account of the nature of possible tremors and the very different behaviour under seismic disturbance of rigid frames and earth buildings. The inexpert introduction of a rigid frame might so transmit energies to a historic structure that collapse would occur where otherwise it might not have done. A broad general rule is to avoid the integration of rigid frames with earth constructions, including those where the only earths are in the mortar.

In consequence of their rigidity and mass, materials such as steel and concrete are rarely appropriate in earth structures as large-scale components. If seismic protection by a framed system is required, it is best that the frame is structure within structure, physically separated in such a way as to allow tolerance for movements which may be anticipated.

Where metals are employed they are generally perforated or expanded sheets of aluminium, zinc, zinc-coated steel or stainless steel. Woven mesh can also be used where a particularly fine texture is required, in which case bronze or brass may be applicable. The most appropriate use of these materials is to improve the connection between dissimilar components, for instance, timber into earth, or to extend areas across which stress is delivered, such as plaster across a crack.

8.6.ii Failure in existing reinforcement

Failure of previous reinforcement in the form of embedded timber members may be corrected where feasible by the replacement with similar timber members proofed against decay or termite attack. This may be given improved cohesion with the linkage of the introduced members to the earth structure by additional reinforcement designed to provide dispersal of stress. Expanded metal lath fixed to replacement timbers may be an effective answer.

Vulnerable zones may be additionally reinforced with the objective of giving longer protection and greater life to the finish. Geotextiles introduced on to the head of a wall or a wall surface may then receive a capping or rendering of natural or amended earth; the objective of the reinforcement is to prevent early cracking or sudden failure in an area which might not be maintained readily or with any certainty. Any such crack would immediately induce a high rate of decay at that

Figure 8.17 A movement joint between two buildings constructed separately. The upper sections were built entirely in earth and the lower of masonry in earth mortars. While the movement joint satisfactorily allows for expansion and contraction, failure to weather it allows water entry (courtesy of Dr Paul Brown).

point. Therefore a form of reinforcement such as a geotextile which will guard against such failure is likely to extend the life of the structure, although it is no substitute for proper repair.

Provision of a non-decaying armature for earths in place of failed vegetable armatures is by definition a method of reinforcement. Timber weaves or palisades against which earth structures have been built may decay through fungal or insect attack even to the point where hardly a fragment of the original material remains. In such cases, where the quality of the panels justifies retention, a poured reinforcement may be considered. If the introduction of water is not a particular problem (as it would be on an internal panel carrying wall paintings on a stucco surface

Figure 8.18 Incompatible structural masses. Two structures built at different times moving differentially which demand the introduction of a slip joint to accept and provide weathering protection for recurrent movement. The repeated repair using mud render is ineffective (courtesy of Rebecca Warren).

Figure 8.19 Timber used as tensile members. The placement of timber horizontally within walls is a common feature even in temperate climates. The behaviour of earth structures frequently depends upon such tensile members. Where through insect or fungal attack the timber ceases to be effective, it may be replaced with proofed timber or other reinforcement. The ability of earth brick to act as a cantilever or corbel is demonstrated at high level (courtesy of Dr Paul Brown).

applied to earths), a grout of modified gypsum plaster injected and reinforced with short chopped fibre could restore the coherence of the panel without disturbance to its surfaces. However, this material would provide very little additional strength. Alternatively, a fluid epoxy grout, which avoids the introduction of water, might be chosen to replace a once rigid armature whereas it would be inappropriately strong as a replacement for earth walling. A danger with such materials is that they may penetrate through the daub with visible results. Other materials which can be introduced in this circumstance include expanding

mixtures of resins designed to foam *in situ*. Such foams may have a relatively short life and they are not reversible.

Structural elements such as vaults may justify the provision of visible tensile members in the form of rod or even wire secured to plates in or on the outer faces of walls receiving thrusts in order to stabilize structural elements which will otherwise fail progressively due to lack of strength in haunchings and abutments. A visible tensile member treated simply and cleanly may well be acceptable in a structure where the need for restraint arises as a result of decay or structural failure.

Figure 8.20 Weald and Downland Open Air Museum). New lath in split oak on chestnut staves replacing a totally decayed panel in a medieval timber frame (architect: the author).

Figure 8.21 Wattle and daub repair. Lime-stabilized daub applied to a panel of second-grade wattle made of split hazel wands on hardwood staves. Application was not simultaneous from both sides.

8.6.iii Concrete

The introduction of liquid concrete, concrete blocks, cement mortars and reinforced concrete components is generally inappropriate in the repair of earth structures because the strength, the mass and the physical characteristics of these materials are not compatible with the earth construction. Perimeter and junction problems arise. Thermal expansion coefficients are different, water absorption and dilation are different and adhesion is poor, with the result that the concrete element is alien and fails to integrate with the structure.

8.6.iv Mesh and fibrous reinforcement

Metallic mesh used as an armature to provide the base for panels and shaped elements of earths will generally be fixed back to a structural framing or masonry mass, providing structural coherence and stiffening. Fixings should always be of the same material as the armature itself to avoid electrolytic action. Since the effects of the use of such framings and armatures are designed on the principle that adhesion is not a calculated part of the strengthening mechanism, bituminous paints may be used, and indeed should be used where there is need for electrolytic insulation and prevention of corrosion.

Glassy filaments are available in different configurations and compositions, derived from matted fibres blown out of melted rock or specially formulated glass. They can be introduced to earth structures in three forms: composite rod, fabric or loose reinforcement.

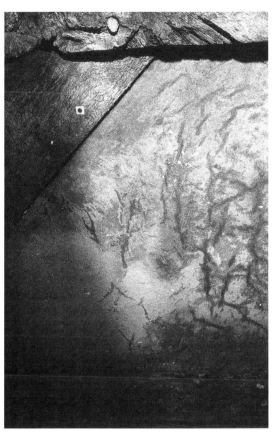

Figure 8.22 Surface texture of newly finished daub coated with limewash. The lack of keying is remedied by integration by the second facing, shortly to be applied.

Figure 8.23 Wattle and daub panel displaying original cracking due to excessively wet application. In an area which had failed completely, cement-based repair has been made successfully. Lime would have been a more appropriate stabilizer.

Resin-bonded carbon fibre might be considered in the category of rod, although its use in earth structures has not yet been reported, and material of such high strength is unlikely to be needed. Glass and carbon-fibre rods have high flexural strengths and may therefore be used in circumstances where bending is an important consideration. They are materials of high strength and, although with appropriate resins they can be bonded, they are not normally formed into mesh, nor woven, nor are they bent. It is possible to protect glass fibres by coating them with synthetic resins and in circumstances where such resins are being used to modify the earths the same or a compatible resin may well be used as a coating if a mat or fabric of fibreglass is

contemplated within the amended earth structure. In general terms it may be taken that a pH value in excess of 9 will damage glasses other than those specifically formulated to resist alkaline environments. Despite the advantages of their being inorganic and, therefore, resistant to biological attack, glass fibres are not conventionally favoured as reinforcement other than as straight resin-bonded rods.

Geotextiles are generally produced from synthetic petroleum-derived plastics, typically polyamides (polymerized diamines), polypropylenes made by polymerization of propylene and polymers of various esters from which fibres are extruded. By compensation-reactions saturated polyesters are produced in which there are no double bonds. These are the basis

Figure 8.24 Mould growth on remade panels of wattle and daub finished with limewash. Basal splash has been allowed to maintain too high a level of dampness.

Figure 8.25 Failure of limewash applied too early to newly completed daub. Unless a daub is thoroughly dry, the limewash, on first application, can overwet the surface. A lamina which has exceeded the liquid limit then underlies the contracting limewash. This substrate is friable and allows the limewash to flake away.

of a range of synthetic fibres, particularly terylene. The general characteristic of all these fibres is strength coupled with flexibility and they are highly durable. As geotextiles they can be supplied as mat or in appropriate loose weave. Being thermo-setting, they can be locked in position on production and by appropriate design of textile a stable material capable of adopting curvilinear profiles can be introduced into an earth structure, providing it with tensile strength along the plane in which it is introduced. By more intimate admixtures improvements in tensile strength in several directions simultaneously can be achieved. Techniques involving this type of application are generally only appropriate where conservation or restoration involves the making good of material which once contained similar reinforcement of a vegetable mixture which will have decayed or disappeared.

Some earth structures depend fundamentally on structures of other materials – none more so than wattle and daub or mud and stud structures in timber frames. Collapse may be irremediable. In such cases, replacement in the original material is essential. No alternative will do because the visual qualities of finished earth daubs are distinctive and vital to the quality of the historic building.

Where possible, earth daubs will be repaired with reconstituted material.

The application of finishes can bring problems. Too early an application of limewash will cause failure. On surfaces particularly where dung has been used and where moist conditions are maintained, algal growth is likely to occur. Historical exactitude may justify this: alternatively, improved detailing is required. In either case a prolonged drying period is desirable before application of the limewash.

9

Case studies

This chapter balances direct practical experience with the social aspects of conservation. A project is described in which construction was directed by structural issues under rigorous conservation policies, and this is counterpoised with conservation in an area where social and emotive imperatives have led to investment in earth buildings for museum purposes; to studies in the regeneration of earth buildings; and to a sumptuous but sympathetic restoration and enlargement.

9.1 Bowhill

Earth structures of all types have been conserved in Britain. One of the most interesting, in the heartland of cob building, is an upper stratum of 'gentry' house which was once rural and is now in the suburbs of Exeter, within an easy ride of the city.

Its significance lies in its sustained care since it (very unusually) came into 'care', i.e. it was acquired by the state. Quite properly, care has been the fundamental criterion in dealing with the property, whose several facets of earth construction make its conservation a crucial exemplar. The italicized sections of the following text are taken, with the generous permission of the architect, Ray Harrison, to whom the author is unreservedly grateful, from his account of the work on Bowhill, published in *Transactions of the Association for Studies in the Conservation of Historic Buildings*, vol. 20, 1995 (courtesy of Stephen Marks, editor) and from his texts prepared for this work.

Grateful acknowledgement is also made to English Heritage for use of material from the forthcoming publication on Bowhill, *English Heritage Research Transactions*, vol. 3, 'Earth',

London, James and James (Science Publishers) Ltd, 1999.

At Bowhill several types of earth construction are integrated with stone masonry and timber framing. The pattern of the work was therefore inherently complex and this complexity was enhanced by a changing philosophical approach. Add to this a pragmatic and empirical approach to the operations which involved local experience, experimental working days with interested participants and managers in training sessions with operatives unfamiliar with their tasks. The approach to materials was in large part traditional both in the use of background knowledge from within the living and documented historic tradition of British (and more specifically, west country) builders in earth and in the 'try it and see' method of selection and improvement. On the face of it this was in direct contrast to the methods of scientific analysis involving understandings of the chemistry, physics and engineering characteristics of the soils, and in contrast to the methodological response where soil blending is designed theoretically and controlled by predetermined parameters. In fact, at management level the exercise was underpinned throughout by an awareness of the science of earth for building but combined with a flexible attitude to controlled empirical experiment allowed by and arising out of the nature of the works contract with directly employed labour.

The restoration of Bowhill is, therefore, an example of a philosophical tracking of the best practice in methodology over a period of transition and of a series of experimental investigations with generous time for analysis and revision of judgement in their execution.

Figure 9.1   Bowhill: work in progress, 1990 (© English Heritage)

Figure 9.2   Bowhill: cross wall first course, 1991 (© English Heritage)

Figure 9.3 Bowhill: Great hall ceiling/Great chamber – applying daub and scat (© English Heritage)

Figure 9.4 Bowhill: blocks (© English Heritage)

In both senses, therefore, Bowhill has been a landmark opportunity in deriving knowledge from the investment of thinking time by a committed and dedicated team able to place principle before commercial imperatives.

Although purchased with the status of a monument, albeit in a seriously neglected state, Bowhill is in character essentially domestic rather than monumental. In it the individual west country techniques of timber framing around crucks combine with earth building and stone masonry to present a range of problems that might almost have been designed as a conservation test-bed.

Bowhill lies in the manor of Barley in St Thomas's Parish a mile west of Exeter. It is on the Exeter-to-Moretonhampstead road – the direct but once hazardous western route out of Exeter to Cornwall via the heights of Dartmoor – and looks back from the east-facing slope known as Dunsford Hill towards the towers of the Cathedral across the River Exe. The building, which dates to around 1500, is best described as a small manor house or mansion. It is typical of late medieval gentry houses in its concern to display status, symbolized by the refinements of a double courtyard and an open hall with, inside, enclosed hearths and chimneys, and traditional features such as screens passage and separate kitchen block.

Although the building has never achieved more than manorial status, Bowhill is important among houses of its type and status in Devon because of the amount of original fabric that survives. Many other parallel cases in the county have suffered more alteration at later extensions and rebuilding. It is outstanding at a national level, because of the timber roof structures of the Hall and Great Chamber. These, though they derive from the 'jointed-cruck' system of the western counties, have a number of special features including a continuous coving above the collar, curved feet to the wind-braces, distinctive intermediate trusses and moulded wall-head purlins.

What already existed before conservation work began, though cloaked in poor and superficial 20th-century trappings and having suffered recent damaging alteration, was a remarkably unaltered example of a late medieval gentry house formed around three sides of a central courtyard, the remaining half of a former double courtyard plan.

The wider objective of the purchase was to place the building in a regional collection of medieval and other minor buildings being put together in the 1970s. At the outset, therefore, there was no direct use to be accommodated in the building and the conservation philosophy was devised accordingly. At this stage the principles were the relatively unaltered methodology of the Ministry of Works approach to monuments in care.

Throughout the 19th and 20th centuries the idea that old buildings should be restored back to their original form has been a constant and revealing feature of the psychology of successive conservation movements. From the desire to restore there arises in turn pressure to remove later additions or alterations felt to be of 'little or no architectural merit or interest', as well as to add back former features to some degree known but missing. In relation to these matters it is important to remember in passing that all judgement is rooted in its period and that the late-20th-century restorer's management of such situations must inevitably differ from that of 100 years ago, though the issues remain the same.

Opposed to restoration and stemming partly from the reform of 19th-century historic building conservation sought by John Ruskin and William Morris, stands another contradictory philosophy. This philosophy stresses the paramount importance of not falsifying the evidence by putting back those things which are no longer there. Morris considered that, where repair to a historic structure is unavoidable, the new work should be identifiable for what it is – later alteration of earlier fabric. He believed that repair should be openly and honestly expressed and, furthermore, that restoration should be avoided as deceitful.

Morris's approach was adopted at the outset by the State Service concerned with preservation, the Ministry of Public Building and Works, and then the Department of the Environment's Directorate of Ancient Monuments and Historic Buildings. In the Report of the Inspector of Ancient Monuments for 1913, 'restoration was regarded as the most heinous offence, making a foreman liable to "instant dismissal"' (q. Thompson, M. W. Ruins, their Preservation and Display, London (B.M.) 1981).

The initial intention was to turn Bowhill into a 'monument' – effectively a national museum piece – once consolidation was complete. Changes over the centuries and the loss of some of the internal partitions caused the first Inspector of Ancient Monuments involved with Bowhill initially to recommend that there was a need to recreate original missing or damaged spaces, and perhaps to make the building more useable as a hybrid monument-cum-exhibition centre.

In line with Morris's dictum, this would need to be done with clearly modern inserted and reversible partitions, which carefully left such evidence as survived clearly visible (Beric Morley to J. R. Harrison).

In the event, however, this particular and very British philosophical line was not held. A change of inspector brought a different approach – restoration.

The building's claim to be a special case, a case for restoration, lay in its relative intactness and quality, but also, as in the first and contrasting conservation strategy, in the didactic, the educational, aims of the organization.

Under the new project team from 1982 restoration became the basic principle. Decisions were then made to recreate certain features, determinable fully or in large part, which were missing or had been transposed by 1976. Decisions to remove some later additions and alterations were also made. The new intention was then virtually to recreate a medieval house; reconstruction and restoration were to be limited to specific cases where basic original arrangements could be confirmed. Otherwise it was rather a question of stripping out what were considered damaging later alterations. The fully roofed nature of the building was a key factor facilitating this line of thinking and action.

Even within this major change in direction, however, restoration was not intended to be total. At the early stages of the work recording was by no means exemplary (in the way the term would be defined today), and it was expected that the original work would be left as visible evidence for interpretation with a minimum of intervention. Over stonework, inside wall finishes were not at first intended. Thus it would be possible at least internally to see the way in which much of the structure had been put together and altered as well as, to some extent, where modern repair, reconstruction or restoration had been carried out.

The works between 1982 and 1987, which involved the conservation and partial reconstruction of the Great Hall wing, were then partly predicated on the restoration process. The building, when conserved and restored, would become a 'monument' (i.e. museum), as is the case to some extent with all buildings in state guardianship, but it would be a museum to a degree 'enhanced' back towards an 'original'.

The museum approach envisaged the presence of custodial staff on the site.

This phase of work preceded the recent revival of interest in earth building and concentrated primarily on the great timber roofs following the long-established principle of working consistently around the building, dealing with each section in toto.

In 1987, with about one-quarter of the building conserved, economic circumstances forced a third change of strategy, bringing a change in ultimate user requirements. With this came the realization that repair of the building's earth walling presented an opportunity for innovative working.

The idea of Bowhill as a custodial site had to be abandoned, largely because of lack of money as English Heritage began to reduce its involvement with unprofitable sites in the late 1980s. It was decided that the building would, if possible, be returned to some practical use rather than stay in English Heritage's ownership on completion. Within this revised context the replacing of certain missing features such as roofing, plastering and partitions would ensure a basic level of convenience for the building's eventual user. For the implementation of most of these important features restoration after the original, rather than the potentially more contrasting 'modern' equivalent, was deliberately chosen. This was an attempt to grasp the nettle of continuing the work in sympathy with the reconstruction already done in the Great Hall in parts of the building where evidence was on the face of it harder to come by. This choice respected and perpetuated the 'didactic' underpinning of the first-stage approach but in a more cautious and self-consciously managed way.

As part of the 1987 review, several other changes of direction were made. First, there was a re-examination of the considerable amount of accumulated documentary research into the building. Secondly, greater emphasis *than before was placed on achieving certainty about original missing arrangements prior to the implementation of repair and reconstruction. Thirdly, there was a renewed concern to complete the archaeological record and analysis of Bowhill.*

An analytical study of this sort involves examination and recording of the building's archaeological context and the application of careful investigative and recording techniques to the fabric as a whole, charting everything in detail and confirming historical development as far as possible. This record begun before and continuing during works contributes to the site archive for the purposes of informing academic study and assisting in the academic interpretation of the building. Such investigations are also one important route to informing proposals that will change a building's fabric since, as has already been emphasized, when action is planned, one should first know what it is that is to be acted upon.

Prior to the introduction of this last strategy the use of earths in the restoration was not driven by total authenticity of material. Lime was used extensively between 1978 and 1987 as a stabilizer. Floors were restored in a cob–lime mixture on the basis of reasonable supposition; all the original evidence had been removed in subsequent improvements. Likewise, cob–lime was used in the end section of the hall, both in new external 'cob' walling and in internal daub panels. These lime–stabilized earths were made with quite large amounts of slaked limes.

Another example of uncertain authenticity under the 1978–87 phase relates to experiments with earth plasters internally.

Experiment with various daub plaster mixes took place, to find a mix suitable for applying internally to cob. All mixes contained some lime and many some dung. Details of mixes were recorded *in situ* on the trial areas, in the kitchen. The evidence suggested that, within the south range at least, internal plasters were originally lime/sand throughout – in fact, virtually identical to the mortar used *on the masonry. While it had always been intended to plaster internally over cob, the decision was eventually taken to plaster over exposed internal stonework also.*

Under the last phase of the works all outstanding internal plastering to cob and

stone was carried out in lime and sand only following the evidence. Under this phase too from 1988 a new cross wall was required, reinstating one which had been built in cob. The choice was between new cob and a distinctively modern material, harking back to Morris and before in the desire to show openly what has been done to the building. It is of course in conflict with the concept of returning a structure to its original architectural character.

In line with the final revised philosophical approach described above, cob was chosen. As well as paying respect to the character of the repairs so far carried out, this choice represented a recognition of the validity of conservation of earths as well as presenting the challenge of reinstating craft skills largely lost to the building industry. Alfred Howard, an elderly local master-craftsman in cob, was invited in with his assistant and controlled experimental work was begun; the skills being developed were passed to another generation and recorded. Eventually works extended from lime/earth floors through stone, timber panel, cob and daub walling to timber frames in floor and roof and new slating and rendering. The 'double-lathed' daub referred to below is quite rare in Britain; it involves putting wet mud into the cavity formed where studs are lathed horizontally on both sides, to produce a solid panel.

During the period 1990–95 encounters with a variety of repair and reinstatement problems set the agenda for the controlled experimental specification, implementation and recording of the use of earth in the works. Two types of record were made – the written and the visual. The written record comprised observations of working processes and subsequent material and (architectural) detail performance. For the visual record, for most of the time, subjects and angles for photography were pre-requisitioned by the architect and inspector; detailed instructions for each shot were given to the professional photographer commissioned to do the work. The intention was eventually to publish an illustrated report on the project. Publication will be achieved in the forthcoming *English Heritage Research Transactions*, vol. 3, 'Earth'.

Controlled experiments with earth at Bowhill (1990–95) extended from the investigation of a number of methods of small-scale raw material and cob block production through largely non-structural, substantial as well as superficial, repair and reinstatement of 500-year-old cob and daub, to new cob walling, to both cob/lime mixes and cob blocks for repair, and to cob blocks for new walling. In addition, wall finishes were examined in some detail. Following that found to have been historically applied, daub plaster was reinstated. This involved the study of the role of animal hair in the mix. Following site precedent, thin lime plasters were trialled and then applied to daub, cob and stone wall sections inside and out, and also to thin ceiling daub. Finally, limewash was attempted straight on to daub.

Limited structural repairs involved tieing timber back through the wall, reinstating door lintels, redirecting load from above failed cob and underpinning a historic feature within a fragment of cob. Major works of structural repair, funded by English Heritage, took place at two sites in the west country, in parallel with those at Bowhill. These were also recorded. Technical reports on the two sites, Bury Barton, Devon and Cullacott, Cornwall (1995, 1996) can be obtained direct from English Heritage.

Although raw materials, which were all of local provenance, were initially used as found, as a result of tests and of experience gained in the course of the work, performance was in some cases subsequently modified by, for instance, the addition of extra aggregate or by alternative methods of trowelling-on. The effects of dryer, and wetter, preparations of the material were examined and conclusions drawn. Some sample testing of the original 500-year-old material took place to enable comparison between that and the repair material. Shrinkage of the clay cob was anticipated and theoretically allowed for in the design of the works in various ways; actual results were then compared with these allowances as the works settled and lost moisture. Techniques for the management of superficial repairs to cob improved as the project proceeded. This particular work consisted of the straightforward 'gobbing up' of holes, filling-in with staged beds one above the other, pre-pegging of the background for mechanical bond, stiff cob–lime mixes, filling with cob brick and daubing-out in thin successive edge-pegged layers. The major cob works enabled the examination of the management of new cob in itself

and in direct association with old, building up in deeper and shallower 'beds' both with and without shutters. The need to reinstate large amounts of missing 'double lath' daub allowed comparison between raw materials appropriate for this purpose and for mass cob walling. The unavoidable introduction of cob blocks provided the opportunity to investigate just how appropriate and convenient these in their turn might be for repair in historic situations.

9.2 ad Dariyeh (also al-Dir'iyeh) Saudi Arabia

This example stands in entire contrast to Bowhill, where detailed concern for principles thrive within a context of committed conservation.

In the helter-skelter modernization of Saudi Arabia its capital city of 40 years ago has vanished. All that remains is the occasional fort or palace. A few kilometres away in an oasis valley, and neglected until the later years of this century, lay the ruins of its predecessor, known as ad Dariyeh, with its citadel township al Turaif.

The conservation of parts of these areas has been carried forward on the impulse of patriotism and heritage, symbolic of a pride in the past, with the advice among others of an American architect, Michael Emrick; also of this author in conjunction with a Saudi architect, Zuhail Faez; and of an Arab architect based in London, Abdulwahid al Wakil, assisted locally by Dr Saleh Lamie.

The importance of this work is less technical than socially interpretative. It has been funded in the impetus to meet the universal urge to retain the built heritage in the face of rampant change. Emrick restored a major palace in al Turaif, together with a mosque and parts of the walls; the author set out principles for the rehabilitation of the extensive settlements of the winding wadi-oasis, and al Wakil renewed a farmhouse as a country retreat for a prince of the line of al Saud. The common thread has been careful research into the visual qualities of the vernacular architecture with a sensitive methodology in its restructuring.

The northern sector of the Arabian peninsula, the Najd, has been permanently if

Figure 9.5 Ad Dariyeh, Saudi Arabia. The Wadi Hanifa is studded with minor fortifications and farmsteads, repaired and retained throughout the 19th and 20th centuries.

sparsely inhabited for a long period and has largely been left to itself throughout history, having little to tempt the invader. Among the few exceptions was the arrival in the early 19th century of the forces of Mohammed Ali, ultimately the usurper of Ottoman power in Egypt, but in 1818 its Governor and deputed to recover for the Sultan the Holy Cities of Mecca and Medina. His purpose in the Najd was to put down the Saud rulers of the Hejaz. At that time their capital was a mud-walled town on a bluff above a low wadi which,

channelling the rainfall from a wide area of upland, retained sufficient water to be a permanent oasis for some several miles. Al Turaif, the walled town, stood above the water course, which was usually dry. It was the animal-powered wells sunk deep into the underlying sandstone that provided irrigation and made the area permanently habitable. Walled defensive points, redoubts rather than castles, were scattered through the Wadi Hanifa. The inhabited area is known as Ad Dariyeh.

The prolonged Egyptian attack was more a series of skirmishes than a siege. Ultimately the superiority of their weaponry gave the Egyptians a victory, after which they razed the town, removed its leadership to humiliation and death in Istanbul and departed.

The Saud dynasty resumed its authority and governed from nearby Riyadh, leaving the Wadi Hanifa to a quiet agricultural existence and the town of al Turaif, with its broken walls and razed buildings, to decay slowly and quietly in the dry climate. Recent concern with historic origins and the pressures of new wealth have brought these quiet gardens and slumbering ruins into the focus of attention with conflicting pressures for reconstruction, conservation and redevelopment.

Several centuries of settled habitation and stability allowed the builders of the Najd and of Ad Dariyeh in particular to develop a style in which the patterning provided by triangulated openings, decorated and moulded string coursing and ribbed surfaces were distinctive. Walls were generally built with lower courses of stone in mud mortar, intermediate levels alternately of stone and brick and in upper levels earth brick or consolidated earth. The entire structure would be rendered over with mud, stabilized in some instances with gypsum, which had the advantage of firing at a lower temperature than lime and, in the dry climate being equally stable. Timber, particularly tamarisk and date palm, was consistently used horizontally over the walls as bearing plates and as stringing members. The mortar used was universally the clay-rich silty sand of the wadi itself and the same material was used for the external render, tempered with gypsum or lime. Wood ash was incorporated in the floors.

The town of al Turaif on its spur above the wadi was never reoccupied after its destruction by the Egyptian army and in the dry climate of the Najd it has remained a ruin field of high jagged walls, pierced with triangular headed openings – a permanent reminder and record of the traditional architecture of the region in the early 19th century.

Subsequent building in Riyadh, prior to the recent spate of development, had evolved somewhat differently in style, although retaining the dogtoothed string coursing and multiplicity of openings typical of Dariyeh. In the

wadi itself, however, architectural development stood still although rural life continued with buildings continually in use and the outlying smaller fortifications being maintained against marauders.

Followers of the puritanical Muslim reformer Muhammad ibn 'Abd al Wahab (1703–92) – the Wahabi sect – had their power base in the wadi in the 19th century. Some of the primitive mosques still survive with low, square towered minarets, reflecting the simple forms of the early years of Islam. The continuity of this community may have been the focus of a social cohesion that maintained the lifestyle and architectural forms of the oasis against the pressures for change.

9.2.i The Omari Palace, al Turaif

Two entirely different problems of conservation were therefore present in a country with more than adequate wealth to deal with them. In al Turaif the options were to conserve as found or to reconstruct: in the wadi the need was to adapt the surviving buildings to continuing use. While little actual conservation has been carried out in al Turaif, there has been significant reconstruction of the city walls and of a number of buildings, including the palace of the ruling dynasty. In the words of Michael Emrick, architect, the mission was to '*maintain, document and, in selective cases, reconstruct significant historic structures at the site in order to interpret the three centuries of history of this site for current and future generations. . . In interpreting a site of this scale and complexity it is easy to help the visitor to transport himself from the culture of the 20th century to the reality of a historic city which was once one of the largest cities in the middle of the Arabian peninsula*'.

The surviving structure possessed sufficient evidence to allow reconstruction which, externally at least, required little speculation other than for the surmounting parapets. The rebuilding involved traditional techniques mechanically assisted, placing stone and earth or mud brick in the manner of earlier construction. Stone foundations were bedded in mud mortar and traditional placing made use of stone fragments laid as string courses or in herringbone lay. The triangular openings were formed using the micaceous, fissiparous

Figure 9.6   Al Turaif, Wadi Hanifa. Saudi Arabia. The gaunt and jagged shapes of the town surviving a century and a half after the sack and demolition.

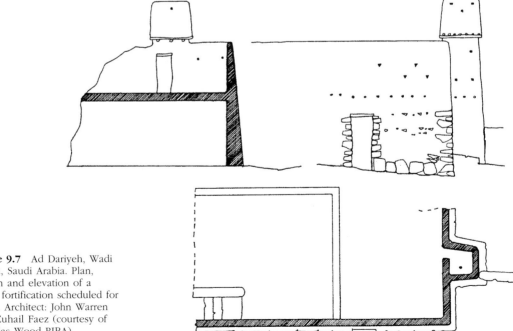

Figure 9.7   Ad Dariyeh, Wadi Hanifa, Saudi Arabia. Plan, section and elevation of a minor fortification scheduled for repair. Architect: John Warren with Zuhail Faez (courtesy of Nicholas Wood RIBA).

Figure 9.8 Palace of Omar Bin Sa'ud, Turaif, Saudi Arabia. Finished state. The rendered walls are topped with lime-washed crenellations.

sandstones of the wadi and the walls were rendered externally with a gypsum mud plaster. Traditionally timbers were often incorporated in the walls as tying members and construction was by post and beam with joisted roofs covered in a mud render. Important elements within the buildings were faced with gypsum plaster, often richly decorated with geometric patterning. Similar decoration was applied to the doors and principal beams; serried ranks of sharp brightly coloured designs in multiple repetition formed the basis of the patterning. In Turaif, however, much of the timber had been removed in the razing of the most important buildings and by later salvage, whereas in the oasis the living communities maintained their establishments, buildings, gardens, farms, mosques and wells.

9.2ii ad Dariyeh

The Wadi Hanifa was coming under increasing pressure from the sprawling adjacent metropolis of Riyadh, when the Saudi government commissioned studies to demonstrate that its environmental qualities need not be eroded in the inevitable changes to the rural economy of the region.

These changes included serious reduction in the water table, growth of road traffic, change from a subsistence economy and pressures of development. The wadi had, for recorded history, been a winding oasis of palm groves, and sheltered gardens where groups of earth-built farmhouses sheltered behind mud brick walls and clustered into small villages. The studies focused on the practical aspects of maintaining unaltered the form and nature of the settlements. To demonstrate the feasibility of retaining these settlements intact it was necessary to show that modern facilities could be incorporated in existing traditional earth buildings without destroying their character and that new ways of using the buildings could relate satisfactorily to their rehabilitation. Careful cost analyses backed up the studies

Figure 9.9 Ad Dariyeh. A traditional domestic coffee hearth in an abandoned farmstead.

and a conservation philosophy with a planning framework was established.

Much of the work involved areas of investigation into traffic, demography and agricultural production which are of no relevance here. Core analyses were devoted to structural techniques, design of traditional features and the perpetuation of a local lifestyle. Methods of increasing the longevity of earth buildings, of repair without destruction and of environmental conservation were explored. Several factors particularly influenced the conclusions: the requirements for vehicle access without adverse impact, the introduction of high-calibre modern services (in particular, bathrooms and cooling/humidification equipment) and the

interior planning which offers Saudi families a lifestyle of privacy and family cohesion.

A crucial part of the advice related to land and building use, primarily devolving on the changing social status of landowners. While land tenure is often a matter of family adherence to its holdings more than of wealth and lifestyle, it remained inevitable that ownerships would change and that wealthier incoming owners would invest in a way which was impossible for those who had derived their living from the harvesting of fruit and vegetables and the herding of goats. Even those who had remained close to the land were destined to change their agricultural methods and increase their expectations.

The problems therefore were complex: without consistent and viable solutions the prospect of retaining the qualities of traditional oasis lifestyle were small and the environment was heading towards ruin by wealth.

Additions to earth structures in concrete block and concrete beam were set aside as unacceptable, and it was a matter of policy that they should be removed or replaced. A defined policy of external amendments only in earths was recognized as fundamental. Roofs, however, were redesigned with cellular semi-structural insulation above the traditional bearing materials – palm, poplar and tamarisk pole joists topped with palm leaf matting. An impermeable membrane was included beneath stabilized earth topped with a tile wearing surface to provide rapid run-off. Rain, though occasionally severe, is always of short duration.

Preformed and prefitted service pods were designed for incorporation into structures by lowering complete units into place; the surrounding voids were filled and foam-sealed *in situ* to provide insulation and minimize inaccessible spaces capable of harbouring vermin. In rebuilding the use of concrete foundation slabs was accepted, the perimeter being built against the normal rough rubble base courses laid into a mud mortar. An underslab membrane was designed to return up the inner face of the wall to slab top but no damp-proof course was introduced into the wall itself. The proposed masonry was of stiff local earth having a 10–20% lateritic clay content placed on the wall and tamped into position. No fibre reinforcement was specified,

(a)

Figure 9.10 (a) Ad Dariyeh. Al
Beijary village, Wadi Hanifa, Saudi
Arabia. Regeneration of a complete
village provides a meaningful
context for the surviving houses,
shown here shadowed but not
shaded. The shaded buildings
replace existing houses or
farmsteads which have been lost,
with the sole exception of the
over-road bridge platform on the
perimeter beside the mosque. The
context of the historic houses is
maintained. Architect: John Warren
with Zuhail Faez.
(b) Plan and section of village
house. A typical village house in
the Wadi Hanifa, Saudi Arabia. The
ground floor plan and cross-section
illustrate a traditional house of the
wadi, in this case an isolated
farmstead. The programme of
rehabilitation involved repair of the
structures themselves and the
introduction of capsules to provide
kitchens and bathrooms. Air
cooling was by remote condensers.
Architect: John Warren with Zuhail
Faez (courtesy of Nicholas Wood
RIBA). *continued*

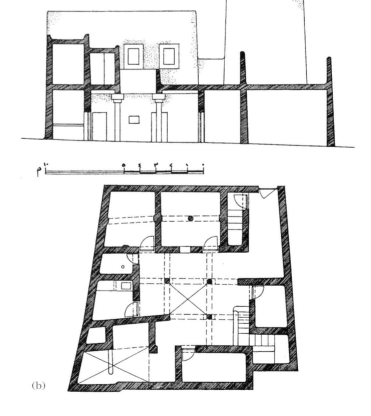

(b)

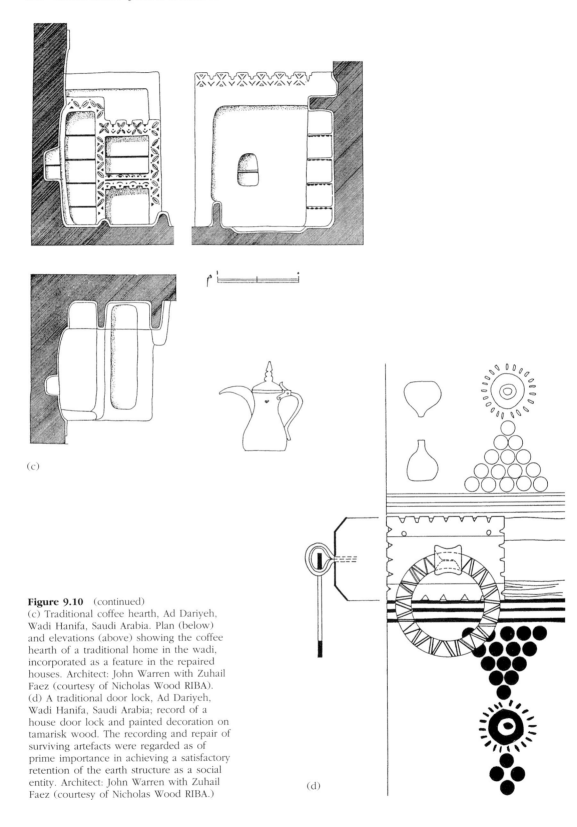

Figure 9.10 (continued)
(c) Traditional coffee hearth, Ad Dariyeh,
Wadi Hanifa, Saudi Arabia. Plan (below)
and elevations (above) showing the coffee
hearth of a traditional home in the wadi,
incorporated as a feature in the repaired
houses. Architect: John Warren with Zuhail
Faez (courtesy of Nicholas Wood RIBA).
(d) A traditional door lock, Ad Dariyeh,
Wadi Hanifa, Saudi Arabia; record of a
house door lock and painted decoration on
tamarisk wood. The recording and repair of
surviving artefacts were regarded as of
prime importance in achieving a satisfactory
retention of the earth structure as a social
entity. Architect: John Warren with Zuhail
Faez (courtesy of Nicholas Wood RIBA.)

(c)

(d)

although the use of straw or grass is common. The intended walls were always thick, sometimes as much as 1 metre and rarely less than 850 mm. On occasion, this technique leads to evident striations in the walls sometimes accentuated by a deliberate overhang repeated in the rendering.

In the detailing of distinctive local features, – rainspouts, parapets, windows – and in the joinery, vernacular building in the wadi was followed with care. Varieties of coffee hearth were explored to give residents the continuity of tradition which is also expressed in the bold simple patterning of colour to be applied to the tamarisk and date palm boarding of doors and cupboards and the soffits of the joists.

Repairs were specified on the basis of simple procedures and practical operations. With the exception of below-ground operations, no cement mortars were specified or permitted. Earths introduced as fillers were to be as close as possible to the shrinkage limit, tamped into position in small parcels. Shrinkage cracks were made good with poured gypsum sealed by mud rendering – a technique possible in dry climates and offering the advantage of expansion on setting. Some earths were specified as being stabilized with gypsum plaster in internal situations and elsewhere lime stabilization was recommended; as much as 10% was proposed for parapets where a lightened colour was advantageous. Fibrous daub was

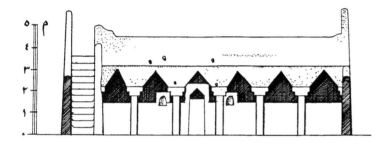

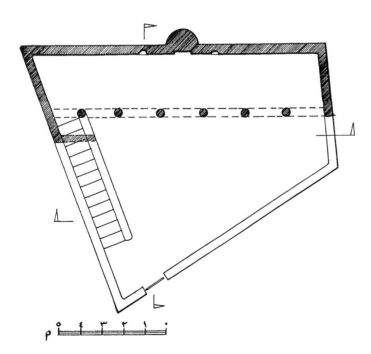

Figure 9.11 Ad Dariyeh. Plan and elevation of the masjid.

specified throughout, even where it was also to be stabilized with lime prior to receiving an external limewash. This technique is applicable by tradition to mosques, masjids and to courtyard internal walls. Within dwellings bright primary colours were recommended in line with established practice.

Even the wells and their lifting gear were studied and drawn as recorded features.

Apart from the typical scale and massing of local buildings the detailing of parapets and roofs, together with texture, window size and patterning and softly contoured walls were established as the critical features of a sympathetic architecture to be introduced in the many circumstances dictated by the adaptation of rural buildings to new and different lifestyles. There was a policy of amendment in a sympathetic manner to provide for new patterns of life.

9.2.iii al-'Udhaibat

With abandonment, many houses fell into decay, as farms went out of production. One such, al 'Udhaibat, was bought by a grandson of King Abdulazid Al-Saud, Prince Sultan, as a site for reconstruction. A sympathy for tradition turned his intentions, with the advice of architect Abdulwahid al Wakil, to undertaking the repair and modification of the building using the techniques and methodology outlined. A master-builder experienced in traditional building methods was available and his presence allowed the architects to dispense with the testing and analytical steps which would otherwise have been essential. Instead they were able to rely on the transmission of practical knowledge and in consequence returned closely to the original construction methods used to build the house. The interior was treated with the powerful simplicity of traditional colours and pattern-making.

The massive social changes in Saudi Arabia have even had their impact on the traditional life of oasis farms and many have fallen into decay. Such was al-'Udhaibat, a once prosperous establishment standing among extensive groves of date palms at a favoured point where the Wadi Hanifa is joined by a tributary. Although in gentle decay, it was still used as a habitation when purchased by Prince Sultan bin Salman bin 'Abd al-Aziz Al Sa'ud.

The Prince responded warmly to its vernacular qualities. It epitomized the architecture of the region and its restoration therefore became a cultural objective. Its rehabilitation would be an important statement in the region. Inevitably the house would move out of one bracket socially into another but this is no more than has happened to many significant examples of housing in the farming and artisan traditions.

The complex consisted of a large traditional well, a small mosque and the house itself with some appurtenances. The west side of the house had fallen away. Prince Sultan steered the project with general advice from the architect, Abd al-Wahid al-Wakil, and engineer Saleh Lamei. The key operative was an 80-year-old master-craftsman whose entire life had been spent, following his father's footsteps, in the building of earth structures in and about Riyadh. Abdullah bin Hamid was effectively the builder and father of the team.

In visual terms the scale, massing and architectural quality of the building were retained, but significant amendments were made to its functioning and structure. It was not, therefore, intended to be a faithful restoration. The objective was a sympathetic re-interpretation in the new circumstances of its use. Structurally it was intended to retain the bulk of the building, although the collapse of the western wall was an early indication of structural weakness, if not inadequacy. This was confirmed by a further internal collapse at an early stage (unfortunately with a fatality) and the thinner walls, unduly thin at 450 mm, were rebuilt to everyone's greater comfort at a thickness of 750 mm. They are battered externally, built in mud brick on stone foundations and rendered with a straw-rich mud. The rebuilding was by no means total and roof structures were retained while sections of walls were rebuilt or packed with additional mud brick at the base to ensure adequate foundation.

Local sources of clays, long exploited, were readily available and, despite experiments in modification with sand, the ultimate conclusion was that the unmodified earths were ideal. Analyses extended through sieving and sedimentation tests to X-ray spectroscopic analysis to determine clay types. Traditionally made mud bricks were moulded and laid out

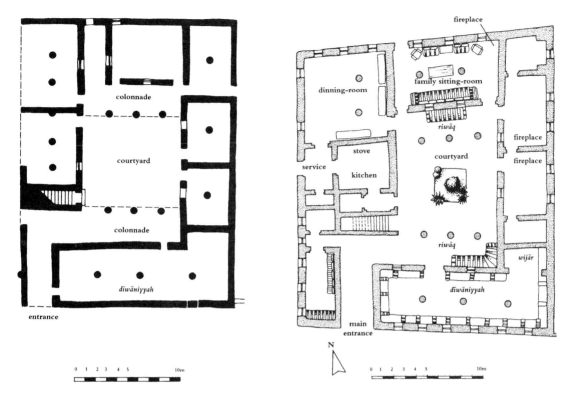

Figure 9.12 Comparitive plans of al-'Udhaibat. The left hand plan shows the house before restoration, the west side having been demolished. The right hand plan shows the house on completion, with the restored west side.

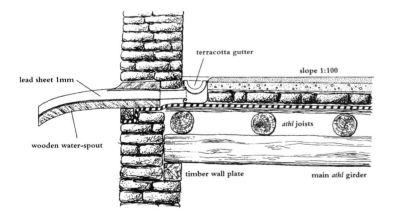

Figure 9.13 Detail of roof construction. Amendments to the traditional system are the introduction of a continuous damp-proof membrane above the palm leaf matting and the introduction of a weak straw-rich mud layer approximately 150 mm deep, over which a clay-rich weathering layer is compressed under a roller. This layer has additional straw in the mix.

on the sandy floor, being turned and set on edge as necessary to provide for even drying. Short chopped-barley straw was used, impregnated against termite attack with a relatively mild organic insect repellent, Dursban 4TC. The local earth proved to have adequate binders of lateritic, kaolinite and attapulgite clays – non-expansive groups. It had a good balance of silt and sand but a rather high salt content, including chlorites and sulphates deriving from the local gypsum strata. However, on the evidence of the success of

Figure 9.14   The courtyard of the building before restoration.

Figure 9.15   The courtyard of the building on completion.

Figure 9.16 Link wall and gate.

other similar operations the material was not washed but potable water was used in the mix in preference to the slightly saline well water. For brick-making 50 kg of chopped straw were used with every cubic metre of semi-dry earth. An equal volume of water was added and the mix was allowed to stand with occasional remixing for some 3 weeks before bricks were moulded in frames, 400 × 200 × 80 mm. The resulting bricks had a compressive strength of almost 4.9 N/mm^2 and on immersion survived 25 minutes before collapse – both very creditable results. The same mix was used for the mortar and the external rendering was similar but with a double quantity of straw. The maturation of the mix is an important part of the process and a visible ferment was produced with a dark stain probably indicative of changes in the straw.

Proofed timber lacing ties were built into the corners of the structure as local tradition wisely suggests. This is always the point where contradirectional stresses cause movement and cracking. Local timbers, tamerisk and athl were used for the roof structures. They were also treated with Durspan.

The chosen roof structure was entirely traditional in form, with a base layer of mud brick over which a dense, compacted layer of clay-rich soil was laid to fall to be surfaced with an even richer weathering layer. Inexpansive clays are the key to success in this type of weathering. There is little surface cracking and hence little penetration initially. It is an important feature that the substrate is a matting of woven palm fronds overlain with a further layer of fronds. This technique of construction has satisfactorily survived intense rainfall; weaknesses proved to be at the junction of roof and parapet walls. The system has shown itself to survive admirably in circumstances where rain, if heavy, is of relatively short duration.

Figure 9.17   The Diwaniyyah.

Figure 9.18 The Riwaq.

Figure 9.19 Boundary walling showing layered construction reflected in the external render.

Figure 9.20   The restored well structure.

One other technical concession was made in the structure in the application of silicon consolidants and inhibitors, Sanotec Befix P and Febersilicon, both of which have proved reasonably effective in the renders and wall cappings.

The interior decoration was undertaken traditionally in juss plastering carved with local patterns and paintwork in the strong vernacular style of the region. Both technically and aesthetically this restoration must be regarded as a significant step in the sensitive reuse of traditional earth buildings.

Although a little grander than it was originally, al-'Udhaibat is one among many historic houses now revered in its enlarged state. The phenomenon of social change driving architectural evolution has been repeated many times and so al-'Udhaibat becomes a fair reflection of the greater changes in society. To freeze such buildings into the constraints of their rural economy is unrealistic and flouts the evolutionary nature of architectural history.

Illustrations for this section on al-'Udhaibat are taken by courtesy of William Facey from the archive created for *Back to Earth, Adobe Building in Saudi Arabia*, published by Al-Turath and the London Centre of Arab Studies, 58–50 Kensington Church Street, London W8 4DB, 1997.

10

Guiding principles

The Venice Charter, written in the decade after World War II, was the product of minds directly concerned with the great stone monuments of the Mediterranean. The savants who gathered to prepare it may well have been conscious that eventually it would be supplemented, taking into account the more subtle aspects of conservation and the more fragile types of construction. In due course this happened and the BURRA Charter has made good this omission, emphasizing the human rather than the monumental aspects and achievements. The burgeoning World Heritage movement has accentuated the need and accelerated the work, giving status to the more fragile and vulnerable elements of building, particularly vernacular architecture, and endorsing the principles which give importance to all significant human works, whatever their method of construction.

With this sea change the conservation movement becomes increasingly concerned with earth structures as robust on the one hand as Silbury Hill and as fragile as wattle and daub or painted clay statuary on the other.

10.1 Authenticity

The philosophies that underlie the conservation of all monuments and vestiges of the work of preceding civilizations remain essentially the same whether the buildings have been in superbly cut marble or in humble earths. Authenticity is the first and prime requisite: the test of authenticity lies both in the material and in form. The stone façades of a cathedral which have been shaved back, redressed and 'improved' may lack authenticity simply because they carry a design whose message is

Figure 10.1 Monumental buildings in earth. The Kasbah Amridil, in the Skoura Oasis, Dades Valley Morocco (courtesy of Jean-louis Michon and the Ministry of Cultural Affairs, Morocco).

Figure 10.2 Repair involves the local community in continuity of its traditions.

of the wilful hand of earlier conservators rather than of the original builders; this, despite the fact that every vestige of stone is original. By contrast, the mud walls of a West African mosque might contain no trace of the surfaces created by its builders some centuries ago. By nature of the construction the surfaces will have been renewed several times since the original building but the form of the construction remains that of the first design. In this sense the mud-built mosque is more authentic than the dressed stones of the cathedral and the comparison demonstrates the fundamental ethic of authenticity.

The conservation movement has now become attuned to this essential criterion and, although there is still much confusion in the lay mind, the principles now established govern professional evaluation and affect understanding of the merits of historic build-ings. Because authenticity lies both in the materials and in the perpetuated shaping it can readily be lost or destroyed and in a sense the loss is irretrievable. A restoration of a damaged or destroyed monument can never carry its original importance and, therefore, never give the same satisfaction as the authentic creation. However, the restoration or recreation may have an important bearing on conservation and is therefore not to be discounted. A build-ing from which a vital component is missing can be restored by the replacement of what is lost. With adequate documentation the restora-tion may be verified as exact. Lacking documentation it may nevertheless be conjec-tural but importantly contributive.

Likewise, in a group of buildings the restora-tion of one missing building may be a similarly valuable component. Therefore there is justifi-cation in the restoration of a part of the struc-ture or even a complete structure which is part of a group. The total authenticity is not dimin-ished and the overall integrity is enhanced.

10.2 Neutrality

Neutrality expresses respect for the historical context of the original objects. It allows a policy by which intervention neither exagger-ates nor diminishes the degree to which a building expresses its origin in time and calibre. There has always been a danger of improvement by the enthusiast as well as depredation by the insensitive. To heighten the historic characteristics or to make a build-ing pretend to a social importance which it never had is as damaging in conservation terms as the degradation that goes with a thoughtless and clumsy work of alteration. Neutrality is a guiding precept which requires that the character of a building is neither enhanced nor degraded in the work of conser-vation and necessary restoration. It is, in effect, an extension of the concept of authenticity.

Usability relates fundamentally to the future of the construction. The preservation of a building is best achieved by its continued use, even if not for its original purpose. Utilization sometimes means alteration and extension and, of all types of construction, earth build-ing offers perhaps the greatest flexibility in reuse for reasons of its general informality and the plasticity of its construction. Alterations which respect these characteristics and build upon them can be adapted to the requirements of new generations without losing their essential qualities. The biggest danger is perhaps the destruction of the evidence of an earlier lifestyle as a result of the adaptation. The conservation objective will therefore be the retention of as much as possible of the existing structure, the appropriate replacement of missing elements and the alteration of the buildings in a way which retains their social status, the character of the materials and the general form which the materials have given to the buildings and to the group.

It is common experience that reuse of earth buildings and their extension is often achieved by additions in other materials. These are frequently inimical to the environmental quality already established. Structures in concrete blockwork with sheet metal roofs, factory-manufactured windows, pipework and paint present a contrast that is inherently destructive of typical earth buildings. The visual damage is complemented by a disparity in usage since building performance varies widely. An earth structure which, through its intrinsic thermal properties, remains habitable in a hot climate may be difficult to use effectively in conjunction with an air-conditioned regime in a structure of more modern materials.

10.3 Intrinsic characteristics

The most profound differentiation between earth structures and those in other types of material is the softness of contour and the frequent battering or inward sloping profile of the walls. These characteristics are shared generally, whereas other distinguishing forms are often regionalized and relate to associated structural techniques. Where timber is not used, or is only used sparingly, as on the Iranian plateau, in sub-Saharan Africa and in the Nile valley, the dome and vault rise characteristically, dominating the architecture and giving form to the groups and internal spaces. Where roofs are formed in timber a different structural method produces a flat-roofed parapet-surrounded building of a type as varied as the tower houses of the Yemen and the flat cityscape of old Baghdad. Where earths are

entirely capped against the weather the buildings take on a form derived from the materials. The low-crouching thatched cottages of the English west country derive their qualities from roofs overhanging massive cob walls, gently curved and moulded to a characteristic profile. Where earths are used as infill panels their framing and the roofing dominate: the panels are subsidiary, contributing an unevenness of surface-modelling which reflects the plasticity of daub and the elasticity of the wattle.

The list could be extended endlessly. The essence is a clear understanding of the qualities that create the character of the building and the skill which allows the conservator to achieve a match in repair and sympathy in restoration.

10.4 The British Standards Guide to the Principles of the Conservation of Historic Buildings

The urge to establish written guidance for conservation activities in buildings has now been carried beyond charters into a guide produced by the British Standards Institution. The need for such a document – seriously contested by some conservators – springs from the pressure for increasing coherence in practice. BS 7913 1998 has been produced by a committee representing the UK bodies concerned with building conservation.

Although in world terms the UK does not possess a great volume of earth construction, it does have a great diversity of materials and methods. It also has a recently discovered enthusiasm for the subject, out of proportion to its place on the world stage. By overviewing the British Standards guide in the context of earth structures it is possible to define a contextual setting for the conservation of earth buildings: this is attempted below. Relevant sections of the guide are italicized where directly quoted and relevant paragraphs are given in parentheses.

The underlying objectives of the Guide are cultural, economic and environmental (1.1).

10.4.i Vernacular and the artisan tradition

In terms of vernacular building it is the local and vernacular materials and construction of buildings of all sorts which *reflect local geology, climate and culture and which contribute to a sense of place* (1.2)

With its wide geological diversity Britain had produced upwards of half a dozen distinct types of building in earths. . .*following variations of topographical, geological, climatic and transport limitations*. . .(1.2). In this diversity earth construction reflects closely the variation in. . .*social and ethnographic factors*. . .(1.2) which the Guide identifies as significant.

In the continuity of the tradition into the industrial era, and the use of mechanised processes of construction for earths. . .the empirical development and refinement of building practice in response to changing needs and circumstances was continued by various skilled trades until the onset of the industrial age and beyond (1.4).

10.4.ii Standards and legislation

In active conservation it is important to apply standards and codes of practice on the basis of principles informed by experience and knowledge including that of relevant legislation. *While the application of particular specifications, structural design codes and calculations can be appropriate in many circumstances, there can be other circumstances where it will be necessary to follow professional experience and judgement* (5.2).

This paragraph alludes to the application of overriding judgements where rules, codes and standards are in mutual conflict or in conflict with the historic building criteria. In earth structures there is often a direct conflict between rules and legislation relating to health, structural strength, permanence and impermeability on the one hand and the performance of earths on the other. Durability has been a particularly difficult matter. Panels of wattle and daub and heavy cob walls alike have been made victim to model bye-law and building regulation requirements on grounds of strength, impermanence and damp-proofing despite it being evident that they have stood successfully, lasted for centuries and, if maintained, been proof against the weather.

The guidance of historic building criteria will go some way to offering welcome support to those concerned with the vernacular tradition.

The Guide emphasizes that. . .British Standards and other specifications and codes of practice should not be applied unthinkingly in the context of building conservation. It is particularly relevant to earth structures that the Guide sets British Standards in a category to be applied with thought and care as to their relevance for the fundamental reason that there may be inherent conflict between standards designed around the control of modern materials and the performance of pre-industrial products.

10.4.iii The process of conservation

Contextually a structure. . .*should always be considered with regard to its history and as a whole* (6.2.1).

The Guide makes clear the need to deal uniformly with the whole of the evidence in a balanced way. In terms of a building which contains elements of earth construction the balance will imbue them ultimately with an appropriate importance.

Caution must be exercised in dealing with the historical component of a building. The specialist nature of this work. . .requires skills that are obtained by specialist conservation training and by experience. . .*These skills include archaeological and survey work and all the activities of conservation* (6.2.2.a). The essence again is a balanced approach in which the practical and cultural objectives are equally valued and where the intrinsic statement made by the building is genuine.

The veracity of a structure depends upon the authenticity of material from any period which is. . .*always of potential value and should be treated with respect* (6.2.2b). Unskilled judgements, particularly in the simple techniques of vernacular building, can result in misleading interpretations.

An initial survey carried out by skilled operatives. . .*to give reasonable confidence that the building, its development and historical importance are sufficiently understood*. . .(6.2.2c) then becomes crucial to success. The retention of evidence is of course fundamental.

Original material should be left undisturbed if possible or it may possibly be separately preserved. . .*the destruction of significant historical evidence without adequate recording is never acceptable* (6.2.2.d & e). In some cases the profiles of earth structures, as found, may

Figure 10.3 Earth structures are profound evidence of settlement patterns (courtesy of Rebecca Warren).

be recorded as a component of the conservation or essential reconstruction as well as being the subject of documentation. For this reason new work should be separable from the old. . .*and should be finished in such a way that it can be distinguished from the original* (6.2.2.f). Since earth structures may decay and can be expected to suffer surface erosion and renewal the incorporation of identifiably dateable materials in renewal is important.

Finally. . .*a proper record of the building and the work done should be made* (6.2.2.g). This admonition, like the requirement to publish archaeological excavations, is too often neglected. The retention of records is noted with the recommendation that in particularly sensitive buildings records are duplicated and archived separately.

It is frequently a very important aspect of vernacular building and earth structures in general that they are profound evidence of human settlement patterns and in this context the guide underlines the need for respect for settlement patterns. In particular. . .*it should be an objective that plot boundaries and the general scale and form of development are maintained* (6.2.3.). Vernacular, and particularly earth structures are frequently part of a cumulative or agglomerative process of building where the shape and pattern of the settlement have an organic arrangement of important subtlety.

The continual use of building stock is emphasized as being a prime objective. Since the greater part of urban architecture is domestic, sensitive reuse is implicitly and

ideally a continuity. . .*alteration should always be kept to a minimum and should, if possible, be reversible* (6.3.2.). Alteration may be necessary because of a change in use or a change in the technology of an established use. The mass of many earth structures may allow accommodation of change while ideally the type of use is retained.

The social advantages of the extended life and use of buildings. . .*which last and continue to be useful, can bring economic benefits in a variety of ways. . .the perception of stability and continuity, confidence and a climate in which economic activity can flourish* (6.3.3.).

10.4.iv Economic factors

The survival or preservation of a building ultimately depends on its continued use and its ability to '*earn its keep*' (6.3.1.). Earth structures which are not maintained decay faster than any other building type. Being less monumental and often humbler than others, failure to encourage or achieve viable uses results in partial or complete loss which in turn may affect others. . .*where poor environment is associated with lack of building maintenance and general decay the buildings can have the potential to contribute very positively. . .conservation can be an effective instrument of economic regeneration* (6.3.4.). The argument is that structures, if maintained, are an asset.

The integrity of groups of vernacular buildings is particularly important. Older buildings. . .*come to symbolize the relationship*

Figure 10.4 Earth can be an important medium for the continuity of vernacular design: in this case the motifs used date back well over a millennium (courtesy of Rebbeca Warren).

between the physical and economic geography of a place, its people and its culture (6.3.6), and in consequence maintain the cohesion of a working community.

There is a broader contemporary aspect to the retention of urban structures because existing buildings. . .*contain embodied energy, derived largely from the labour invested in them when they were built which is dissipated and lost when they are destroyed. . .they are also inherently maintainable and almost entirely benign in terms of toxicity* (6.4.2).

This is more fundamentally true in earths than in any other form of construction. The manufacturing techniques are of the simplest kind, with minimal energy input and minimal transport costs.

10.4.v Environmental objectives

Taking a world view perhaps no architectural type exceeds vernacular building in earth in. . .*establishing or enhancing the visual environment. . .in a way which contributes more effectively to a sense of community. . . they are familiar and are a known and recognizable feature of a place* (6.4.3). In terms of local economy the continuity of traditional

methods may have important implications in terms of social cohesion and lifestyle. . . *Settlements of all sorts are complex organisms which can react to surgery in unpredictable ways. The continued use and reuse of buildings, if repairs and other work are well executed, will almost always make a positive contribution to the local environment* (6.4.3).

The guide underlines the need for logical processes of thinking and in particular an analytical balance of objectives before the detail of work is planned. . .*it is always desirable to identify and balance objectives in principle before attempting to resolve issues which can arise in practice* (6.5).

10.4.vi Conservation in practice

The overall approach to conservation of traditional building. . .*should be considered as a whole and treated in a holistic way* (7.1.1). It is particularly true of earth structures that. . .materials and method of construction and patterns of air and moisture movement should be properly understood.

The Guide advocates a cautious and minimalist approach. All significant work should be preceded by thorough documentary research

Figure 10.5 The castle at Bam in Iran maintained traditionally with a mud and straw mix by two men and a donkey working from the mixing base in the compound near the centre of the picture.

and physical investigation. Where possible work should be reversible (7.1.1). Documentation in the field of earth structures is often minimal and available archive material may be confined to a few photographs or nothing at all. Research is occasionally surprisingly rewarding and a sound knowledge of what has been lost is fundamental to restoration.

Good conservation involves. . .*a presumption against restoration* (7.1.2). It is important to understand and work with the fabric of the building, not against it, and to be flexible and imaginative.

Since earth buildings are inherently fragile although very long lasting, maintenance should be. . .planned as a regular routine. . .as building conservation becomes more science-based so an understanding of the relevant treatments and processes increases in importance (7.1.3.).

Particularly with earth structures a conservative approach to repair is important. . .*replacing decayed material on a like for like basis is preferred, although there are occasions when it is more appropriate to use non-traditional materials and methods if these are more*

Figure 10.6 Traditional materials for rebuilding counter the ubiquitous advance of the concrete block (courtesy of Dr Paul Brown).

discreet and allow more existing fabric to remain in situ undisturbed (7.3.1). Almost uniquely in the construction process it is possible, with earths, to reconstitute the bulk of the original material for repair.

Where restoration is to be considered it may be demanded by. . .*a lacuna or void in an otherwise complete or coherent design. . .or by a functional, structural or constructural reason for [replacing] the missing element* (7.3.2.2).

Where restoration is essential and justifiable. . .*new material introduced in the course of like for like repair and restoration should match the original materials as closely as possible* (7.3.2.4). Matching should not target replication of the existing colour, texture or appearance but should be specified to weather towards the character of the fabric. . .*different materials chosen to match at the outset will. . .match less well as they age* (7.3.2.4.). Many forms of earth construction accept repair with like materials entirely naturally and invisibly. It is in the nature of exposed earths to weather quickly and harmoniously. Alien materials, however, often stand out in stark contrast.

Where new and old work is to be juxtaposed. . .*new work should not damage, mask or devalue the old either physically or visually.* It should be of appropriate quality and should complement the old. It should be reversible. . .*and should combine to form a composite building or group of buildings of overall architectural and visual integrity* (7.4.2).

This *de minimis* overview seeks to highlight singificant principles set out in the British Standards guide and serves to link its contents with the conservation of earth structures; but those concerned with applying the skills of conservation will need to study the guide in full and in the context of the environment and country in which they are working. The British Standards Institution cannot be held responsible for shortcomings due to the brevity of these extracts and it must be emphasized that the achievement of the guide itself is a compromise between the many reservations of the contributing organizations and individuals. In seeking to promote good practice it embraces current philosophies in the field.

It is the essence of environmental conservation to rely upon a common denominator – the consensus view of what is important and is to be extracted from the past by the present to be retained for the future. Particularly in the context of earth structures the ultimate tenet of conservation is that it is better to retain than to destroy, for destruction is irretrievable.

Acknowledgement

Extracts from BS 7913: 1998 are reproduced with the permission of British Standards Institute under licence number PD/1998 0859. Complete editions of the standards can be obtained by post from British Standards Institute Customer Services, 389 Chiswick High Road, London W4 4AL.

Index

Conservation of Brick

John Warren

- Technical and scientific background is clearly defined
- Current conservation philosophies are integrated with practise
- The only current work on this subject
- Especially important – the conservation of fired clays

John Warren's invaluable book describes historic brick and terracotta, setting out the causes of failure and decay, analysing available materials and evaluating processes of repair and applicable conservation philosophies. It provides the conservator, owner and student of building conservation with a comprehensive resource.

Brickwork, with tile and terracotta, is one of those materials so universal, so apparentaly permanent and so much part of our everyday lives that its conservation is presumed to be understood. This is very largely untrue. Most brickwork is cursorily maintained andoften subject to serious abuse. Neglect and clumsy repair are all too frequent, and the really skilful repair based on a full understanding of the mechanisms of decay is all too rare.

This book is a companion volume to John Warren's *Conservation of Earth Structures*, also part of the Butterworth-Heinemann conservation series.

CONTENTS:
An outline of the evolution of bricks and brickwork; Brickmakers, bricks and brickworks; Historic brick; Refined brick earths – terracottas and tiles; Behaviour of fired clay materials; Response to failure; Physical repair; Destructive processes in brickwork; Biodeterioration; Natural and man-made pollution; Protection, consolidation and repair; Mortars, renderings and plasters; Case studies; Good practice; Checklists for conservation activities in brickwork; Bibliography; Index.

1999, 304 pp, 19 line drawings, 154 black and white photographs, 27 colour plates
246 X 189 mm, Hardback, 0 7506 3091 4, £65.00

Historic Floors Their History and Conservation

Editor/Author Jane Fawcett MBE Hon FRIBA LRAM ARCM Grad Dipl Cons A.A Architectural Historian and Historic Buildings Consultant

- The first book in the UK to be devoted to historic floors
- In association with ICOMOS UK (International Council on Monuments and Sites)
- Contributions from leading practitioners in their field

Historic Floors introduces an important and largely neglected subject and considers conservation methods in a European context.

It traces the history of some of the great floors of Europe from the fourth century BC and outlines the development of mosaic, tiles, marble and parquetry floors in secular buildings. The early Christian pavements in basilicas, temples and cathedrals, the creation of medieval tiles, ledger stones and monumental brasses, their destruction by iconoclasts and re-creation during the Gothic Revival, are also discussed.

Leading authorities, archaeologists, architects and archivists consider the latest methods of recording and repairing cathedral floors, including those of cathedrals, country houses, the monumental tiled pavements of the Palace of Westminster and other public buildings.

CONTENTS INCLUDE: The archaeology of church and cathedral floors; the retrieval of damaged inscriptions; the conservation of excavated floors; the repair of tiled and mosaic pavements; problems of country house parquetry and marble floors; the management of visitors.

May 1998, 260 pp,
over 80 colour photographs,
246 x 189 mm, Hardback, 0 7506 2765 4, £60.00

Laser Cleaning in Conservation

Edited by Martin Cooper
National Museums and Galleries on Merseyside, Conservation Centre, Liverpool

- Introduction by John Larson
- The first book to be written on this subject
- Describes pioneering work of John Larson, Martin Cooper and colleagues
- Includes chapter on laser cleaning of paintings
- In-depth and wide-ranging case studies show methods and results of use of lasers in conservation.

This pioneering book will give anyone with an interest in the conservation of artworks a basic understanding of the laser cleaning technique.

The author maintains that improved knowledge of the technique will lead to more widespread and responsible use of lasers. in the conservation of artworks.

CONTENTS: Introduction; Basic principles of a laser; Removal of surface layers by laser radiation; Practical laser cleaning with a Q-switched Nd: YAG laser; Case studies, Lasers in the conservation of painted artworks; Case studies; Future developments; Appendix; Index.

January 1998, 112 pp, 43 colour photographs, 38 line illustrations, 11 photographs, 276 x 219 mm, Hardback, 0 7506 3117 1, £30.00